W9-BZC-603

STATISTICAL MODELS FOR CAUSAL ANALYSIS

STATISTICAL MODELS FOR CAUSAL ANALYSIS

ROBERT D. RETHERFORD
MINJA KIM CHOE
Program on Population
East–West Center
Honolulu, Hawaii

A Wiley-Interscience Publication
JOHN WILEY & SONS, INC.
New York • Chichester • Brisbane • Toronto • Singapore

This text is printed on acid-free paper.

Copyright © 1993 by John Wiley & Sons, Inc.

All rights reserved. Published simultaneously in Canada.

Reproduction or translation of any part of this work beyond that permitted by Section 107 or 108 of the 1976 United States Copyright Act without the permission of the copyright owner is unlawful. Requests for permission or further information should be addressed to the Permissions Department, John Wiley & Sons, Inc., 605 Third Avenue, New York, NY 10158-0012.

Library of Congress Cataloging in Publication Data:

Retherford, Robert D.
Statistical models for causal analysis/Robert D. Retherford, Minja Kim Choe.
p. cm.
Includes bibliographical references and index.
ISBN 0-471-55802-8 (cloth)
1. Multivariate analysis. I. Choe, Minja Kim. II. Title.
QA278.R47 1993
519.5′35—dc20 93-23423
CIP

Printed in the United States of America

10 9 8 7 6 5 4 3 2 1

PREFACE

This book provides a quick overview of statistical models commonly used in causal analyses of nonexperimental data in the social and biomedical sciences. Topics covered are simple bivariate regression, multiple regression, multiple classification analysis, path analysis, logit regression, multinomial logit regression, and survival models (proportional hazard models and hazard models with time dependence).

The goal is to impart an intuitive understanding and working knowledge of these models with a minimum of mathematics. The strategy is to simplify the treatment of statistical inference and to focus primarily on how to specify and interpret models in the context of testing causal theories. General principles and procedures of statistical inference and model testing are covered, but derivations of sampling distributions and standard errors are omitted, thereby circumventing a great deal of complex mathematics. On the other hand, a fair amount of mathematical detail is retained in the discussion of model specification and interpretation.

It is assumed that the reader knows elementary algebra and has mastered the equivalent of a graduate-level introductory course in social or biomedical statistics. Calculus is occasionally used, mainly in illustrative derivations involving minimization or maximization (e.g., minimizing a sum of squares or maximizing a likelihood function) and in the development of survival models in Chapters 7–9.

The computer programs by means of which illustrative examples were produced are from BMDP, LIMDEP, SAS, and SPSS, because these major statistical computing packages tend to be generally available. For each type of example, at least one and usually two computer programs are provided in Appendix A of this book. Whenever possible, one of these is from LIMDEP, in order to make things easier for the reader who is new to computers and wishes to learn how to use only one package. The programs may be implemented either on a mainframe computer or on a microcomputer. Some models, such as hazard models with time dependence, may take considerable time to execute on a microcomputer and are best done on a mainframe computer.

The examples are taken from the 1974 Fiji Fertility Survey, which was part of the World Fertility Survey. The individual-record tape from the Fiji Fertility Survey is in the public domain and provides a convenient source of constructed examples. Although the examples are taken from the field of population studies, the methods are generally applicable to a wide variety of problems in the social and biomedical sciences as well as other fields.

Social science and biomedical professionals wishing to upgrade their methodological skills will find this book useful. The book is also suitable for use as either a core or a supplementary textbook for courses in social or biomedical statistics, following an introductory course. Normally these would be graduate-level courses. However, because of the highly simplified treatment, the book may sometimes be appropriate for advanced undergraduate courses as well. We have used the book as the textbook for a one-semester graduate course, with good results.

ACKNOWLEDGMENTS

We are grateful to John Bauer, S. I. Laskar, Norman Luther, Andrew Mason, and Mahmudul Haq Khan for helpful comments and suggestions, and to Robin Loomis for assistance in preparing graphs and figures.

Robert D. Retherford
Minja Kim Choe

Honolulu, Hawaii

CONTENTS

1 Bivariate Linear Regression **1**

1.1. Terminology, 1
1.2. Fitting a Least-Squares Line, 2
1.3. The Least-Squares Line as a Causal Model, 5
1.4. The Bivariate Linear Regression Model as a Statistical Model, 7
1.4.1. Simplifying Assumptions, 8
1.5. Statistical Inference: Generalizing from Sample to Underlying Population, 10
1.5.1. Hypothesis Testing, 12
1.5.2. Confidence Intervals, 15
1.5.3. t Values and Z Values, 16
1.5.4. p Value, 17
1.5.5. Importance of a Good Spread of Values of the Predictor Variable, 18
1.5.6. Beware of Outliers!, 20
1.5.7. Beware of Selection on the Response Variable!, 21
1.5.8. Presentation of Results, 22
1.6. Goodness of Fit, 23
1.6.1. Standard Error of the Estimate, s, 23
1.6.2. Coefficient of Determination, r^2, and Correlation Coefficient, r, 24
1.7. Further Reading, 28

2 Multiple Regression **29**

2.1. The Problem of Bias in Bivariate Linear Regression, 30
2.2. Multiple Regression with Two Predictor Variables, 31
2.3. Multiple Regression with Three or More Predictor Variables, 34
2.4. Dummy Variables to Represent Categorical Variables, 34
2.4.1. Categorical Variables with Two Categories, 34
2.4.2. Categorical Variables with More Than Two Categories, 36
2.5. Multicollinearity, 38
2.6. Interaction, 40
2.6.1. Model Specification, 40
2.6.2. More Complicated Interactions, 43
2.6.3. Correlation without Interaction, 45
2.6.4. Interaction without Correlation, 45
2.7. Nonlinearities, 46
2.7.1. Quadratic Specification, 46
2.7.2. Dummy Variable Specification, 49
2.8. Goodness of Fit, 51
2.8.1. Standard Error of the Estimate, s, 51
2.8.2. Coefficient of Determination, R^2, and Multiple Correlation Coefficient, R, 52
2.8.3. Corrected R^2 and Corrected R, 53
2.8.4. Partial Correlation Coefficient, 54
2.9. Statistical Inference, 55
2.9.1. Hypothesis Testing, Confidence Intervals, and p Values for a Single Regression Coefficient, 55
2.9.2. Testing the Difference Between Two Regression Coefficients, β_i and β_j, 56
2.9.3. Testing Effects When There Is Interaction, 57
2.9.4. Testing Effects When There Is a Nonlinearity, 58
2.9.5. The ANOVA Table, 59
2.9.6. The Omnibus F Test of the Hypothesis $\beta_1 = \beta_2 = \cdots = \beta_k = 0$, 60
2.9.7. Test of the Hypothesis That Some of the β_j Are Zero, 61
2.10. Stepwise Regression, 62
2.11. Illustrative Examples, 63
2.11.1. Example 1, 63
2.11.2. Example 2, 66
2.12. Further Reading, 68

3 Multiple Classification Analysis **69**

3.1. The Basic MCA Table, 70
3.1.1. Unadjusted Values, 70
3.1.2. Adjusted Values, 73
3.1.3. Unadjusted and Adjusted, *R*, 75
3.1.4. A Numerical Example, 76
3.2. The MCA Table in Deviation Form, 78
3.2.1. First Approach to Table Set-up, 78
3.2.2. Second Approach to Table Set-up, 78
3.2.3. A Numerical Example, 81
3.3. MCA with Interactions, 82
3.3.1. Table Set-up, 82
3.3.2. A Numerical Example, 84
3.4. MCA with Additional Quantitative Control Variables, 85
3.4.1. Table Set-up, 86
3.4.2. A Numerical Example, 87
3.5. Expressing Results from Ordinary Multiple Regression in an MCA Format (all Variables Quantitative), 88
3.5.1. Table Set-up, 88
3.5.2. A Numerical Example, 89
3.5.3. Why Present Ordinary Regression Results in an MCA Format?, 89
3.6. Presenting MCA Results Graphically, 90
3.7. Further Reading, 92

4 Path Analysis **93**

4.1. Path Diagrams and Path Coefficients, 93
4.2. Path Models with More Than One Exogenous Variable, 100
4.3. Path Models with Control Variables, 105
4.4. Saturated and Unsaturated Path Models, 107
4.5. Path Analysis with Standardized Variables, 109
4.5.1. Standardized Variables and Standardized Path Coefficients, 109
4.5.2. Path Models in Standardized Form, 110
4.5.3. Standardized Versus Unstandardized Coefficients, 113
4.6. Path Models with Interactions and Nonlinearities, 114
4.7. Further Reading, 118

5 Logit Regression **119**

5.1. The Linear Probability Model, 119

5.2. The Logit Regression Model, 121
5.2.1. The Logistic Function, 122
5.2.2. The Multivariate Logistic Function, 125
5.2.3. The Odds and the Logit of *P*, 126
5.2.4. Logit Regression Coefficients as Measures of Effect on Logit *P*, 128
5.2.5. Odds Ratios as Measures of Effect on the Odds, 129
5.2.6. The Effect on the Odds When the Predictor Variable Is Categorical with More Than Two Categories, 132
5.2.7. The Effect of the Predictor Variables on the Risk *P* Itself, 134
5.2.8. Interactions, 135
5.2.9. Nonlinearities, 136
5.3. Statistical Inference, 137
5.3.1. Tests of Coefficients, 137
5.3.2. Testing the Difference Between Two Models, 138
5.4. Goodness of Fit, 140
5.5. MCA Adapted to Logit Regression, 142
5.5.1. An Illustrative Example, 142
5.5.2. Table Set-up and Numerical Results, 144
5.6. Fitting the Logit Regression Model, 147
5.6.1. The Maximum Likelihood Method, 147
5.6.2. Derivation of the Likelihood Function, 147
5.7. Some Limitations of the Logit Regression Model, 150
5.8. Further Reading, 150

6 Multinomial Logit Regression **151**
6.1. From Logit to Multinomial Logit, 151
6.1.1. The Basic Form of the Multinomial Logit Model, 151
6.1.2. Interpretation of Coefficients, 153
6.1.3. Presentation of Results in a Multiple Classification Analysis (MCA) Format, 153
6.1.4. Statistical Inference, 157
6.1.5. Goodness of Fit, 157
6.1.6. Changing the Reference Category of the Response Variable, 157
6.2. Multinomial Logit Models with Interactions and Nonlinearities, 159

6.3. A More General Formulation of the Multinomial Logit Model, 160
6.4. Reconceptualizing Contraceptive Method Choice as a Two-Step Process, 163
6.5. Further Reading, 165

7 Survival Models, Part 1: Life Tables **167**
7.1. Actuarial Life Table, 168
7.1.1. Statistical Inference, 173
7.2. Product-Limit Life Table, 175
7.2.1. Testing the Difference Between Two Survival Curves, 177
7.3. The Life Table in Continuous Form, 178
7.4. Further Reading, 180

8 Survival Models, Part 2: Proportional Hazard Models **181**
8.1. Basic Form of the Proportional Hazard Model, 182
8.1.1. A Simple Example, 183
8.1.2. Addition of a Second Predictor Variable, 184
8.1.3. Redefinition of the Baseline Hazard, 185
8.1.4. Relative Risks as Measures of Effect on the Hazard, 187
8.1.5. Hazard Regression Coefficients as Measures of Effect on the Log Hazard, 188
8.1.6. The Effect on the Hazard When the Predictor Variable Is Categorical with More Than Two Categories, 190
8.1.7. Interactions, 191
8.1.8. Nonlinearities, 192
8.1.9. Changing the Reference Category of a Categorical Predictor Variable, 194
8.2. Calculation of Life Tables from the Proportional Hazard Model, 194
8.2.1. The Hazard Model in Survivorship Form, 194
8.2.2. The Problem of Unobserved Heterogeneity, 195
8.2.3. Test for Proportionality, 196
8.3. Statistical Inference and Goodness of Fit, 198
8.4. A Numerical Example, 198
8.5. Multiple Classification Analysis (MCA) Adapted to Proportional Hazard Regression, 205
8.6. Further Reading, 205

9 Survival Models, Part 3: Hazard Models with Time Dependence **207**

9.1. Time-Dependent Predictor Variables, 207
 9.1.1. Coefficients and Effects, 208
 9.1.2. Choosing a Baseline Hazard Function, 209
 9.1.3. MCA Adapted to Hazard Regression with Time-Dependent Predictor Variables, 210
9.2. Time-Dependent Coefficients, 214
 9.2.1. Coefficients and Effects, 214
 9.2.2. Survival Functions, 216
 9.2.3. MCA Adapted to Hazard Regression with Time-Dependent Coefficients, 218
9.3. Further Reading, 220

Appendix A Sample Computer Programs **221**

A.1. Description of the FIJIDATA File, 221
A.2. SAS Mainframe Programs, 223
 Program 1. SAS Program for Multiple Regression (Program Used for Table 2.2), 223
 Program 2. SAS Program for Partial Correlation Coefficient (Program Used for Table 3.4, Adjusted R Column), 224
 Program 3. SAS Program for Logit Regression (Program Used for Table 5.1), 225
 Program 4. SAS Program for Multinomial Logit Regression (Program Used for Table 6.4), 226
A.3. Preparing Data for Survival Analysis, 227
 Program 5. SAS Program to Create Input Data File for BMDP and LIMDEP Programs for Life Table and Hazard Model Analysis of Progression from Fifth Birth to Sixth Birth, 228
A.4. BMDP Mainframe Programs, 230
 Program 6. BMDP1L Program for Actuarial Life Table: Progression from Fifth Birth to Sixth Birth

for Fijians and Indians (Program Used for Table 7.1 and the Actuarial Life Tables in Table 7.2), 230

Program 7. BMDP1L Program for Product-Limit Life Table: Progression from Fifth Birth to Sixth Birth for Fijians and Indians (Programs Used for Product-Limit Life Tables in Table 7.2), 232

Program 8. BMDP2L Program for Proportional Hazard Model: Progression from Fifth to Sixth Birth for Indians (Program Used for Table 8.1), 233

Program 9. BMDP2L Program for Hazard Model with Time-Dependent Covariate (Program Used for Table 9.2), 235

Program 10. BMDP2L Program for Hazard Model with Time-Dependent Covariate and Time-Dependent Effect (Program Used for Table 9.3), 237

A.5. LIMDEP Programs for IBM-Compatible Personal Computers, 238

Program 11. LIMDEP Program for Multiple Regression (Program used for Table 2.2), 238

Program 12. LIMDEP Program for Partial Correlation Coefficient (Program Used for Table 3.4, Adjusted R column), 239

Program 13. LIMDEP Program for Logit Regression (Program Used for Table 5.1), 240

Program 14. LIMDEP Program for Multinomial Logit Regression (Program Used for Table 6.4), 241

Program 15. LIMDEP Program for Life Table (Actuarial and Product-Limit): Progression from Fifth Birth to Sixth Birth, Indians (Program Used for Table 7.2), 242

Problem 16. LIMDEP Program for Proportional Hazard Model: Progression from Fifth Birth to Sixth Birth, Indians (Programs Used for Table 8.1), 243

Problem 17. LIMDEP Program for Hazard Model with Time-Dependent Covariate: Progression from Fifth Birth to Sixth Birth, Indians (Program Used for Table 9.2), 243

Problem 18. LIMDEP Program for Hazard Model with Time-Dependent Covariate and Time-Dependent Effect: Progression from Fifth Birth to Sixth Birth, Indians (Program Used for Table 9.3), 244

Appendix B Statistical Reference Tables **245**

References **251**

Index **254**

1

BIVARIATE LINEAR REGRESSION

Multivariate analysis encompasses a variety of statistical methods used to analyze measurements on two or more variables. *Regression analysis*, which is the subject of this book, is a major subset of multivariate analysis that includes methods for predicting values of one or more response variables from one or more predictor variables. The simplest form of regression analysis is *bivariate linear regression*, involving a straight-line relationship between one response variable and one predictor variable. The discussion of bivariate linear regression in this chapter provides the conceptual building blocks for more complex forms of regression analysis presented in later chapters.

1.1. TERMINOLOGY

Suppose that two variables X and Y are related approximately as $Y = a + bX$. Y is called the *response variable*, and X is called the *predictor variable*. Alternatively, Y is called the *dependent variable*, and X is called the *independent variable*. If it is appropriate to think of X as determined independently of the equation and to think of Y as explained (at least in part) by the equation, then X is often called an *exogenous variable* and Y an *endogenous variable*; in this case, X is also called an *explanatory variable*. More generally, X is also referred to as a *regressor*. When we fit the line, $Y = a + bX$, to data, we speak of *regressing Y on X*.

Variables may be *quantitative* (e.g., income) or *qualitative* (e.g., urban–rural residence). Quantitative variables may be either *discrete* (e.g., number of children ever born) or *continuous* (e.g., time). Qualitative variables are also called *categorical variables* or *factors*. In regression analysis, all variables, whether quantitative or qualitative, are represented numerically. Qualitative variables are represented by dummy variables, which will be discussed later.

We shall often refer to an equation, such as $Y = a + bX$, as a model. Generally speaking, a *model* is "a description of the relationships connecting the variables of interest. . . . The process of model-building consists of putting together a set of formal expressions of these relationships to the point when the behaviour of the model adequately mimics the behaviour of the system" (Kendall and O'Muircheartaigh, 1977). In this book, models are always stated in the form of equations, including any assumptions necessary for model fitting and statistical inference. The choice of mathematical form of the model (including choice of variables and the statement of underlying assumptions) is referred to as *model specification*. The term *specification error* is used to indicate an incorrect model specification.

1.2. FITTING A LEAST-SQUARES LINE

By way of hypothetical illustration, Figure 1.1 shows a fitted straight-line relationship, with the response and predictor variables defined as

Y: number of children ever born (fertility)
X: number of completed years of education

The plotted points might represent, say, ever-married 35 to 49-year-old women in a fertility survey. Each point represents a woman in the sample. Only a few of the sample points are shown.

The idea is to fit a line of the form $Y = a + bX$ through the points (but not necessarily coinciding with the points). The quantities a and b must be estimated. The term a is called the *constant term* or *intercept* (i.e., Y intercept) of the equation, and b is called the *coefficient* of X. b is also called the *slope* of the regression line. Once the values of a and b are determined, the equation of the line offers a means both for summarizing the relationship between fertility and education and for predicting fertility from education.

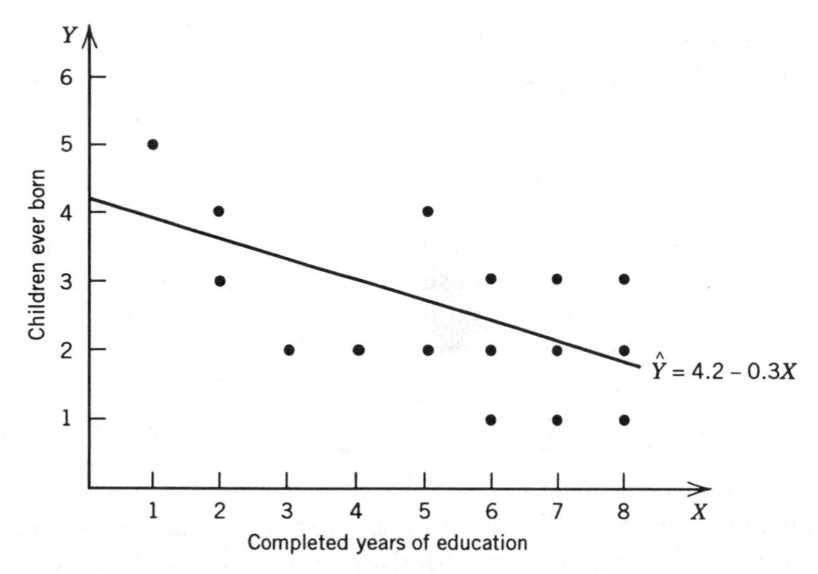

Figure 1.1. Regression of fertility on education from hypothetical sample data.

Suppose that we have a sample of n women. Then there are n *sample points* (X_i, Y_i), where X denotes completed years of education, Y denotes number of children every born, and subscript i denotes the ith woman in the sample ($i = 1, 2, \ldots, n$). The equation of the line to be fitted is

$$\boxed{\hat{Y}_i = a + bX_i} \tag{1.1}$$

where $\hat{Y}_i$ (pronounced Y-hat-sub-i) denotes the *predicted* value of Y obtained from the regression equation by substituting in the *observed* value of X for the ith women. Observed values of X and Y are denoted by uppercase letters without a hat.

Usually Y_i and $\hat{Y}_i$ will not be the same. We define a *residual*, e_i, as the difference

$$e_i \equiv Y_i - \hat{Y}_i \tag{1.2}$$

(The symbol "$\equiv$" means "equal by definition".) If the point (X_i, Y_i) is above the line, e_i is positive, and if the point is below the line, e_i is negative.

It follows from (1.2) that $\hat{Y}_i = Y_i - e_i$. By substituting this expression into (1.1), we can write (1.1) alternatively as

$$\boxed{Y_i = a + bX_i + e_i} \tag{1.3}$$

Although there are many ways to fit the line in (1.3), the method of *ordinary least squares* (OLS) is usually preferred because of its ease of computation and desirable statistical properties. To fit the least-squares line, we first form the sum of squared residuals, S:

$$S = \sum_{i=1}^{n} e_i^2 = \sum_{i=1}^{n} \left(Y_i - \hat{Y}_i\right)^2 = \sum_{i=1}^{n} [Y_i - (a + bX_i)]^2 \tag{1.4}$$

We choose a and b so that this sum of squared residuals is minimized. We do this by taking the partial derivative of S first with respect to a and second with respect to b, and then equating each derivative separately to zero, yielding two equations in two unknowns, a and b. These two equations are called the *normal equations*, which can be written as

$$an + b\Sigma X = \Sigma Y \tag{1.5}$$

$$a\Sigma X + b\Sigma X^2 = \Sigma XY \tag{1.6}$$

For notational simplicity, the limits of summation, $i = 1$ to n, and the i subscripts of X and Y are dropped from these two equations.

Equations (1.5) and (1.6) may be solved for a and b to yield

$$\boxed{b = \frac{n\Sigma XY - \Sigma X \Sigma Y}{n\Sigma X^2 - (\Sigma X)^2}} \tag{1.7}$$

$$\boxed{a = \frac{1}{n}(\Sigma Y - b\Sigma X)} \tag{1.8}$$

Fitting the line amounts to calculating a and b by these formulae.

Equation (1.8) may be rewritten as

$$a = \overline{Y} - b\overline{X} \tag{1.9}$$

where $\overline{Y} = (\Sigma Y)/n$ and $\overline{X} = (\Sigma X)/n$, the sample means of Y and X, respectively. Equation (1.9) can be rewritten in turn as

$$\bar{Y} = a + b\bar{X} \tag{1.10}$$

From this equation it is evident that the point $(\bar{X}, \bar{Y})$ *always falls on the fitted least-squares line.*

1.3. THE LEAST-SQUARES LINE AS A CAUSAL MODEL

Suppose that we have a *theory* (loosely speaking, our verbal notions of what causes what) that says that education is a cause of fertility (i.e., variation in X is a cause of variation in Y), and suppose also that the model $\hat{Y} = a + bX$ adequately portrays this causal relationship. Actually this is not a good model of the causal relationship between education and fertility, but let us suppose for the moment that it is in order to illustrate some basic ideas. In Chapters 2 and 4 we shall consider better models that allow for multiple causes and take into account indirect effects as well as the direct effect of education on fertility.

If we accept the equation $\hat{Y} = a + bX$ *as a causal model, the coefficient* b *can be interpreted as a measure of the effect of* X *(education) on* Y *(children ever born).* This interpretation may be clarified by considering what happens when X increases by one unit. Let us denote the new predicted value of Y, after a one-unit increase in X, by $\hat{Y}^*$:

$$\hat{Y}^* = a + b(X + 1) = a + bX + b = \hat{Y} + b \tag{1.11}$$

When X increases by one unit, Y increases by b units. Thus b is the increase in Y per unit increase in X. *In other words,* b *is the effect on* Y *of a one-unit increase in* X.

The increase, b, in (1.11) is added to $\hat{Y}$, and in this sense the simple bivariate regression model is an *additive model*. The model is considered additive even if b is negative. (In later chapters we shall consider some multiplicative models, where the effect of increasing X by one unit is to multiply $\hat{Y}$ by a factor.)

A hypothetical result of this fitting procedure might be something like $\hat{Y} = 4.2 - 0.3X$, as illustrated in Figure 1.1. This means that women with no education ($X = 0$) have a predicted family size of 4.2 children. The value of 4.2 corresponds to the *Y intercept* of the graph in Figure 1.1, which is the value of $\hat{Y}$ where the line crosses the Y axis at $X = 0$. The coefficient of X is -0.3, meaning that each additional year of education reduces the number of children ever born by 0.3 child. Thus a woman with 2 years of education has a predicted family size of $4.2 - (0.3)(2) = 3.6$ children.

Note from equation (1.11) that the effect of X on Y does not depend on the value of X. No matter what the starting value of X is, the effect on Y of a one-unit increase in X is always b. This means, as in the above example, that an increase of two units in X would increase $\hat{Y}$ by $b + b = 2b$ units, again demonstrating the additive nature of the model.

Although we have been treating the equation $\hat{Y} = a + bX$ as a causal model, in fact causality is often very difficult to establish. In experimental studies, X can be manipulated, and other sources of variation in Y can be controlled. In this circumstance it is reasonable to assert that the change in Y is caused by the manipulation of X. Assertion of causality is much more problematic in observational studies, where designed experiments are not feasible. In this case, the plausibility of a causal relationship is increased if there is a clear time ordering of the variables, but even then the relationship between X and Y may be due to the effect of some third, uncontrolled variable. Moreover, the researcher is often confronted by a fuzzy time ordering, especially in cross-sectional survey data. The question of whether a causal interpretation is plausible cannot be answered from statistics alone. It must ultimately be answered on the basis of theory (Stuart and Ord, 1991, pp. 967–968).

So far we have assumed one-way causation, from X to Y. But what if Y is also a cause of X, so that there is two-way causation? Of course, if we think of cause as temporally prior to effect, then in theory there can be no such thing as two-way causation. In practice, however, two-way causation is a common problem that occurs when the researcher cannot specify variables and time as finely as he or she would like. Hanushek and Jackson (1977, p. 15) provide an example of two-way causation involving the reciprocal effects of school attendance and school achievement. The problem is that the measures of X (school attendance) and Y (school achievement) from the sample survey under consideration represent cumulations or averages over time. It is because of the fuzziness of the time dimension that it is possible for causation to run both ways. Two-way causation is commonly referred to as *simultaneity*, although this term also has a broader meaning in the context of simultaneous equation systems (not covered in this book, except for the simple path models in Chapter 4).

If two-way causation is present, the interpretation of b *is ambiguous. We can no longer interpret* b *as the effect of a one-unit increase in* X *on*

Y, *because the effect of* Y *on* X *is mixed in with it*. It must be emphasized, however, that the bivariate regression model (or the other regression models considered in this book) itself cannot tell us whether two-way or even one-way causation is present. The model provides a way of summarizing the covariation between X and Y, *but the mere fact of covariation is not sufficient to establish or prove causation*.

Of course, a least-squares line can be fitted even when we don't have a causal model. For example, we could regress life expectancy (Y) on the crude death rate (X) for a sample of nations and fit a line through the points. In this case, it would not make sense to interpret b as the causal effect of the crude death rate on life expectancy. However, it is quite legitimate to think of the crude death rate as a *predictor* of life expectancy. A regression of life expectancy on crude death rate might be useful for predicting life expectancy for nations where, because of data limitations, estimates of the crude death rate are available but estimates of life expectancy are not.

1.4. THE BIVARIATE LINEAR REGRESSION MODEL AS A STATISTICAL MODEL

As already mentioned, ordinary least squares (OLS) is not the only method by which one may fit a line through a scatter of points (X_i, Y_i). The other possible methods need not concern us here. Suffice it to say that when certain reasonable assumptions are made, the least-squares line has desirable mathematical properties that make it attractive as a statistical model.

A model becomes a *statistical model* when it is fitted to sample data with the aim of generalizing beyond the sample to the underlying population from which the sample was drawn. The generalizations about the underlying population are *probabilistic*. For example, one might conclude that there is a 95 percent chance, or probability, that the true effect, β, of education on fertility in the underlying population is somewhere in the range $b \pm 0.2$, where b is the estimated effect derived from the sample. (In this context β is thought of as fixed, whereas b is variable because it depends on the sample drawn.)

Underlying population characteristics, when expressed in numerical form, are called *parameters*, and sample characteristics are called *statistics*. For example, in the previous paragraph β is a parameter and b is a statistic. b is also called an *estimator* of β.

The reader should note that some texts use $\hat{\beta}$ instead of b to denote the estimator of β. In this book we use lowercase Greek letters to denote population regression coefficients, and corresponding lowercase Roman letters to denote sample estimates of regression coefficients.

1.4.1. Simplifying Assumptions

The bivariate regression model becomes a mathematically tractable linear statistical model under certain simplifying assumptions about the underlying population from which the sample is drawn. These assumptions, which are illustrated in Figure 1.2, are as follows:

1. *Linearity.* To begin with, it is assumed that the underlying population is large, so that there are many Y values at each X. Let X_i denote a particular value of X, and let Y_i denote a variable that ranges over the many Y values at $X = X_i$. (Note that these definitions

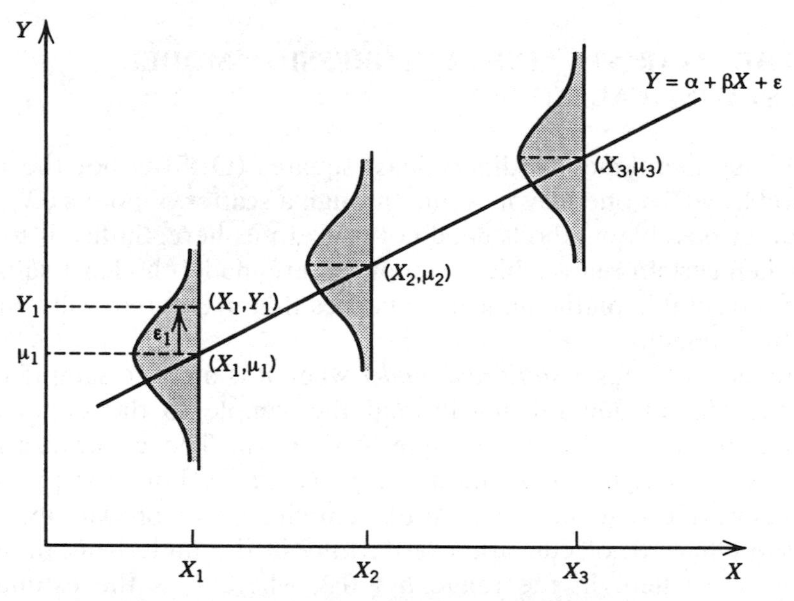

Figure 1.2. Regression of fertility on education in the underlying population: illustration of simplifying assumptions needed for statistical inference. **Note:** Y is number of children ever born, and X is number of completed years of education. For the illustrative point, (X_1, Y_1), $Y_1 = \mu_1 + \varepsilon_1$. The shaded areas represent the distribution of Y at each X. The means, μ_i, of these distributions, fall on the regression line, and the variance of Y, which measures the spread of the distribution, is the same at each X. The distributions need not be normal, although they are drawn as such.

of X_i and Y_i differ from our previous definition, where (X_i, Y_i) denoted the values of X and Y for the ith individual in the sample.) Let μ_i denote the mean of the variable Y_i at $X = X_i$. *Then, by assumption, the μ_i lie on a straight line, known as the true (population) regression line*:

$$\mu_i = \alpha + \beta X_i \tag{1.12}$$

Define

$$\varepsilon_i \equiv Y_i - \mu_i \tag{1.13}$$

The variable ε_i, which is unobserved, is called the *error* or *disturbance*. Substituting (1.12) into (1.13) and rearranging, we get

$$Y_i = \alpha + \beta X_i + \varepsilon_i \tag{1.14}$$

If, for a given value of $X = X_i$, we take the population mean of both sides of (1.13), we obtain $\mu_i = \mu_i + E(\varepsilon_i)$, or $E(\varepsilon_i) = \mu_i - \mu_i = 0$, where the operator E denotes "expected value of" or mean. Therefore, an equivalent statement of the linearity assumption is that (1.14) holds, and that $E(\varepsilon_i) = 0$ at every $X = X_i$.

2. *Homoscedasticity.* This "equal variance" assumption means that, in the underlying population, the variance of the variable Y_i, denoted by σ^2, is the same at each $X = X_i$. Equivalently, the variance of ε_i is σ^2 at each $X = X_i$. (This equivalence is easily proved: For any particular $X = X_i$, $E(\varepsilon_i) = 0$ and $E(Y_i) = \mu_i$. Therefore, $\sigma^2_{\varepsilon_i} = E[(\varepsilon_i - E(\varepsilon_i))^2] = E(\varepsilon_i^2) = E[(Y_i - \mu_i)^2] = \sigma^2_{Y_i} = \sigma^2$.)

3. *Independence.* The error terms ε_i are statistically independent.

We call $Y_i = \alpha + \beta X_i + \varepsilon_i$ the *underlying population model* and call $Y_i = a + bX_i + e_i$ (or $\hat{Y}_i = a + bX_i$) the *estimated model*.

An important part of the definition of a linear statistical model is the procedure by which the X's and the Y's are sampled. The X's may be selected at random, by purposeful selection, or by some combination of the two (such as stratified random sampling or systematic sampling). The requirement for sampling of the Y's, however, is more stringent. For any given X in the sample, a Y value must be selected at random from the population distribution of Y values at that particular value of X.

In experimental studies, sampling may be thought of in the following way: Treatments (X) are purposeful or fixed in advance, whereas the experimental outcomes (Y) may be thought of as selected randomly from a distribution of possible outcomes in which random error comes into play. In social science, on the other hand, the data usually come in the form of sample surveys. In this case, we assume in this book that the type of sampling is random sampling, where both X and Y are selected randomly.[1]

If the assumption of linearity is valid, and if the sampling procedures are as just described, then it can be shown that the OLS estimators a and b, in the sample regression $\hat{Y} = a + bX$, are *unbiased estimators* of the underlying population parameters α and β. By "unbiased" we mean that, over repeated samples from the same population, $E(a) = \alpha$, and $E(b) = \beta$. If the assumptions of homoscedasticity and independence are also valid, the standard errors of a and b have simple estimation formulae.

In reality, of course, the assumptions of linearity, homoscedasticity, and independence are almost always violated to some extent. Then the OLS estimators, a and b, may be biased ($E(a) \neq \alpha$, $E(b) \neq \beta$), and tests of statistical significance (discussed later) may be invalid. In many applications, however, it has been found that the OLS estimators are *robust* to violations of the assumptions. By "robust" we mean that most of the time the estimators and statistical tests are not greatly affected by violations of the assumptions. Later we shall discuss briefly some unusually serious violations that the analyst should beware of.

1.5. STATISTICAL INFERENCE: GENERALIZING FROM SAMPLE TO UNDERLYING POPULATION

Let us consider first the regression coefficient, b, from the estimated least-squares line, $\hat{Y} = a + bX$. How can we use b to make statements about β, the coefficient of X in the underlying population from which the sample was drawn?

The first step is to consider the *sampling distribution* of b, which is the probability distribution (relative frequency distribution) of b over

[1]Most large-scale sample surveys employ multistage cluster sampling. In this case the sampling of the Y's is not strictly random so that the usual formulae for standard errors of coefficients, which assume random sampling of the Y's, constitute downwardly biased estimators. Because most statistical computing packages use formulae based on random sampling of the Y's, the standard errors printed by these packages are also downwardly biased. Adjustment for the "design effect" of clustering can be made, but a discussion of these adjustments is beyond the scope of this book. For further discussion, see Cochrane (1963).

repeated samples from the same underlying population. The sampling distribution of b is derived theoretically, not by actually taking repeated samples. Normally we have only one real sample available for analysis.

Even though we have only one sample, statistical theory enables us to estimate the standard deviation of the sampling distribution of b. This estimated standard deviation, called the *standard error* of b, is denoted s_b. s_b can be calculated from the sample data and is printed out, along with b itself, by standard computer programs for linear regression. In this book we shall be content with understanding the meaning of s_b, without going into the mathematics of how it is derived.

If a value of β is hypothesized and s_b has been calculated from the sample data, b may be transformed into a *t statistic*:

$$\boxed{t = \frac{b - \beta}{s_b}} \tag{1.15}$$

where t is the difference between b and β, measured in standard deviation units. The hypothesis according to which a value is assigned to β is called the *null hypothesis*. The reason why b is transformed to t is that, under conditions described in the next paragraph, t has a known sampling distribution, called *Student's t distribution*, with desirable statistical properties.

If ε_i is normally distributed for every $X = X_i$, the sampling distribution of $(b - \beta)/s_b$ conforms to the t distribution, even in small samples.[2] If ε_i is not normally distributed, statistical inference from small samples is problematic. In large samples, the sampling distribution of $(b - \beta)/s_b$ conforms approximately to the t distribution even when ε_i is not normally distributed. In this case, the t distribution itself converges toward a normal distribution.

In this book, when we consider small samples (less than about 120 observations, as discussed shortly), we shall assume normally distributed errors. When we consider larger samples, we will not invoke this assumption, which is then unnecessary.

[2]If the errors are normally distributed, it can also be shown that the maximum likelihood estimates of α and β are identical to the OLS estimates, but we shall not pursue this topic in this book. The maximum likelihood approach to estimation is discussed briefly in Chapter 5 in the context of logistic regression.

Associated with t is a quantity called *degrees of freedom*, abbreviated d.f. *In* (1.15) *there are* $n - 2$ *degrees of freedom associated with the computed value of* t, *where* n *denotes the sample size.* A full understanding of the concept of degrees of freedom requires a fairly sophisticated knowledge of mathematical statistics. The point to be appreciated here is that one needs to know the number of degrees of freedom associated with t in order to ascertain the sampling distribution of t, and we shall illustrate this later with an example. One need not fully understand the concept of degrees of freedom to make intelligent use of the t statistic.

The *t distribution* (Table B.1 in Appendix B) is somewhat more spread out than the *normal distribution* (Table B.2 in Appendix B). However, if the sample size is moderately large (different cutoffs are used, but a convenient rule of thumb is d.f. $= n - 2 > 120$), the t distribution is approximately normal.

The t distribution is symmetrical about zero. The same is true of the normal distribution. (See the graphs at the top of Tables B.1 and B.2 in Appendix B.)

Regression programs in such statistical computing packages as SPSS, SAS, BMDP, and LIMDEP routinely print out values of b, s_b, and, under the null hypothesis that $\beta = 0, t$. In this case, (1.15) simplifies to $t = b/s_b$. A null hypothesis that $\beta = 0$ is a commonly used hypothesis that the predictor variable has no effect on the response variable. *If, on the other hand, the null hypothesis is that* β *equals some value other than* 0, *the printed value of t must be ignored, and t must be recalculated from* (1.15) *as* $t = (b - \beta)/s_b$. The values of b and s_b used in the calculation of t are the same in either case.

1.5.1. Hypothesis Testing

Classical hypothesis testing begins with a hypothesis, called the *null hypothesis*, about the value of the coefficient of X in the underlying population. This hypothesis is customarily denoted as H_0:

$$H_0: \quad \beta = \beta_0 \tag{1.16}$$

As already mentioned, β_0 is usually taken to be zero, the hypothesis then being that the predictor variable has no effect on the response variable. (As before, our use of the word "effect" in this context presupposes that we are testing a hypothesis about the influence of X on Y.)

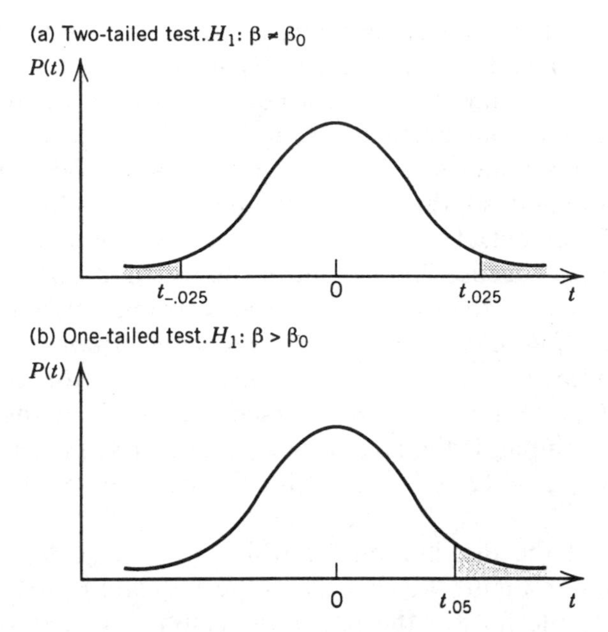

Figure 1.3. Illustrations of hypothesis testing at the 5 percent level of significance. **Note:** $P(t)$ denotes the probability distribution of t. If the sample value of t (based on H_0: $\beta = \beta_0$) falls in the shaded region (called the *rejection region*), then H_0 is rejected. H_1 is accepted, but the probability of erroneously rejecting H_0 may be as high as 5 percent, depending on the sample value of t.

The null hypothesis H_0 is tested against an *alternative hypothesis* H_1. Often H_1 is the hypothesis that we really believe in, and H_0 is a "straw man" that we expect will be rejected as a result of the test. For example, when H_0 states that there is no effect ($\beta_0 = 0$), we often expect to show that there actually is an effect. But we start by hypothesizing that there is no effect.

If we do not have strong theoretical notions *a priori* that either $\beta < \beta_0$ or $\beta > \beta_0$, then we test H_0 against the alternative hypothesis

$$H_1: \quad \beta \neq \beta_0 \tag{1.17}$$

This alternative is called a *two-sided alternative*, and the test is called a *two-tailed test*.

In choosing between H_0 and H_1, we look at the theoretically derived sampling distribution of t under the assumption that H_0 is true, as illustrated in Figure 1.3a. If the null hypothesis is true, the mean of the sampling distribution of t [where $t = (b - \beta_0)/s_b$] is 0. If our sample value of t is close to 0, we have no reason to reject H_0.

Only a value of t for our particular sample that is far from 0 can convince us to reject H_0 and accept H_1 instead.

But how far is "far"? The most commonly used definition is that the value of t for our particular sample must be extreme enough to fall outside a symmetrical interval about $t = 0$, the boundaries of which are defined so that the probability of t falling outside the interval is 5 percent. The values of t at the boundaries are denoted $-t_{.025}$ and $t_{.025}$, because 2.5 percent of the distribution is beyond each boundary in one or the other tail of the distribution, as illustrated in Figure 1.3*a*. *The value of* $t_{.025}$ *is not computed from the data but is looked up instead in the table of t critical points (also called critical values) in Table B.*1 *in Appendix B* (see the graph at the top of the table). For example, if the size of our particular sample is $n = 42$, so that d.f. $= n - 2 = 42 - 2 = 40$, it is seen from this table that $t_{.025} = 2.021$.

The part of the distribution for which $t < -t_{.025}$ or $t > t_{.025}$ (the shaded region in Figure 1.3*a*) is called the *rejection region*. If t for our particular sample falls in the rejection region, it means that such an extreme value of t would occur less than 5 percent of the time (over repeated samples from the same underlying population) if H_0 were true. This probability is small enough that we reject H_0 and opt instead for H_1. In doing so, we run a risk, which may be as high as 5 percent, of erroneously rejecting H_0.

If t for our particular sample falls in the rejection region in Figure 1.3a, we say that *b differs significantly from* β_0 *at the* 5 *percent level of significance*.

Although 5 percent is the most commonly used cutoff for significance testing, 10 percent, 1 percent, and 0.1 percent are also used. Other values are also possible but are not commonly used. Clearly the choice of, say, 1 percent instead of 5 percent would make it more difficult to reject H_0. Thus a 1 percent level of significance is said to be a "higher" level of significance than a 5 percent level of significance, even though $1 < 5$.

If we have reason to believe that β is actually greater than β_0, then the alternative hypothesis, H_1, is

$$H_1: \quad \beta > \beta_0 \tag{1.18}$$

The alternative is called a *one-sided alternative*, and the test is called a *one-tailed test*, illustrated in Figure 1.3b. In this case, negative values of t do not invalidate the null hypothesis, H_0: $\beta = \beta_0$. Only values of t that are high enough in the upper tail can convince us to reject H_0 and opt for H_1 instead. The rejection region is now the shaded region for which $t \geq t_{.05}$. Again we look up the cutoff value of t, in this case $t_{.05}$, in Table B.1 in Appendix B. For example, if the size of our particular sample is $n = 42$, so that d.f. $= n - 2 = 42 - 2 = 40$, it is seen from this table that $t_{.05} = 1.684$. If $t > 1.684$ for our particular sample, we conclude that such an extreme value would occur less than 5 percent of the time if H_0 were true, so we opt for H_1 instead.

If the alternative hypothesis is H_1: $\beta < \beta_0$, we proceed similarly. The only difference is that the rejection region is now in the lower tail instead of the upper tail of the distribution.

For either a one-tailed or a two-tailed test, the *level of significance* may be defined more generally as the probability of erroneously rejecting H_0 when opting for H_1. This means that the "higher" the level of significance is, the less likely we are to make a mistake when opting for H_1.

In classical hypothesis testing, the level of significance is chosen in advance, before H_0 is tested against H_1. Then, in the test, the difference $b - \beta_0$ is found to be either significant or nonsignificant, depending on whether t falls in the rejection region.

If the sample size is expanded, s_b tends to become smaller, $t = (b - \beta_0)/s_b$ tends to become larger, and the likelihood, or probability, that $b - \beta_0$ is statistically significant at a specified level increases.

1.5.2. Confidence Intervals

From the definition of the t distribution, we have that

$$\Pr(-t_{.025} < t < t_{.025}) = 0.95 \tag{1.19}$$

where Pr denotes probability. Allowing that β may be nonzero, and substituting $(b - \beta)/s_b$ for t, we can rewrite (1.19) as

$$\Pr\left(-t_{.025} < \frac{b - \beta}{s_b} < t_{.025}\right) = 0.95 \tag{1.20}$$

If we multiply the string of inequalities within parentheses by s_b, subtract b, multiply by -1 (thereby reversing the direction of the inequalities), and then flip the string of inequalities around, (1.20) becomes

$$\boxed{\Pr(b - t_{.025}s_b < \beta < b + t_{.025}s_b) = 0.95} \tag{1.21}$$

Equivalently,

The 95 percent *confidence interval* for β is $b \pm t_{.025} s_b$	(1.22)

where b is the estimate of β derived from our particular sample. Equation (1.21) and statement (1.22) say that there is a 95 percent probability (i.e., we are 95 percent confident) that the interval includes β. Again $t_{.025}$ is not computed from the data but is looked up in Table B.2 (Appendix B) with d.f. $= n - 2$.

The 95 percent confidence interval may be viewed as an *interval estimate* of β. The size of the interval is a measure of the precision of the estimate. The sample estimate b is a *point estimate* of β.

If d.f. $= n - 2 > 120$, t is distributed approximately normally, as mentioned earlier, so that $t_{.025}$ is close to 1.96. (See Tables B.1 and B.2 in Appendix B.) Then the 95 percent confidence interval for β can be calculated as $b \pm 1.96 s_b$.

Confidence intervals for other levels of confidence (say 99 percent or 99.9 percent) are constructed analogously. The only difference is that $t_{.005}$ or $t_{.0005}$ is used in place of $t_{.025}$.

We may use the confidence interval to test the null hypothesis, H_0: $\beta = \beta_0$. If β_0 falls in the confidence interval, we do not reject H_0. If β_0 does not fall in the confidence interval, we reject H_0 and opt for H_1: $\beta \neq \beta_0$ instead. *This use of the confidence interval for testing H_0 is equivalent to the two-tailed test described in the previous section.* Confidence intervals are not normally used for one-tailed tests, although one could define a confidence interval that is open on one end.

1.5.3. *t* Values and *Z* Values

For d.f. $= n - 2 > 120$, the t distribution approximates a standard normal distribution, with a mean of 0 and a standard deviation of 1. The larger the sample, the closer the approximation becomes. The t value then approximates a standard normal deviate, which is conventionally referred to as a *Z value* or *Z score*. Z values are calculated from the data in the same way as t values, as $Z = (b - \beta)/s_b$.

When hypothesis testing is based on the normal approximation, we use the table of standard normal probabilities (Table B.2 in Appendix B) in place of the table of t critical points (Table B.1 in Appendix B). This is advantageous because the normal distribution is simpler to use than the t distribution, insofar as the normal distribution is not tabulated by degrees of freedom.

In the t table (Table B.1 in Appendix B), a few selected probabilities are given in the column labels, and corresponding t values (called *critical points* or *critical values*) are given in the body of the table. In the normal table (Table B.2 in Appendix B), the mode of presentation is reversed: Z values are given in the row and column labels, and probabilities are given in the body of the table. Because degrees of freedom need not be specified in the normal table, it is possible to present a detailed sequence of probabilities rather than a few selected probabilities.

In the normal table, the Z value corresponding to a given cell entry is obtained by adding the row label and the column label for that cell. For example, the Z value corresponding to the probability .4052 is 0.2 (the row label) plus 0.04 (the column label), or 0.24.

The last row of the t table, labeled ∞ (corresponding to an infinitely large sample size), provides t values that are identical to Z values. Consider, for example, the cell in the t table located at the intersection of the shaded row and column. The t value is 1.96, corresponding to an upper-tail probability of .025. Because the sample is infinitely large, 1.96 may be treated as a Z score. Now consider the normal table. The cell with row label 1.9 and column label .06 contains .0250. Thus the two tables correspond.

As another example, consider the t value of 1.645 in the last row of the t table, corresponding to an upper-tail probability of .05. Now consider the normal table. The cell corresponding to $Z = 1.64$ shows a probability of .0505, and the cell corresponding to $Z = 1.65$ shows a probability of .0495. Linear interpolation of these two sets of numbers yields a probability of .0500 corresponding to $Z = 1.645$, so that once again the t table and the normal table correspond.

1.5.4. *p* Value

A disadvantage of choosing in advance a level of significance of, say, 5 percent is that t may be close or far from the cutoff, but the degree of closeness is not indicated. An alternative is to present the *p value*, also called the *prob value* or *observed level of significance*. The p value is the test level that would just barely allow rejection of H_0, given the t value calculated from the sample. It is the probability of observing a t value at least as extreme as the t value calculated from the sample, under the assumption that H_0 is true.

p values may also be one-tailed or two-tailed. Two-tailed p values, based on the hypothesis H_0: $\beta = 0$, are often printed out by standard computer programs for regression. One can easily convert a two-tailed p value into a one-tailed p value: If t falls in the half of the sampling distribution that includes the one-tailed rejection region, the one-tailed p is obtained by dividing the two-tailed p by 2. If t falls in the other half of the sampling distribution, the one-tailed p is obtained by dividing the two-tailed p by 2 and subtracting the result from 1.

If one is testing H_0: $\beta = \beta_0$, where $\beta_0 \neq 0$, the p value as well as the t value printed out by standard computer programs for regression must be ignored, because they both assume that $\beta = 0$. Instead, one must calculate t as $t = (b - \beta_0)/s_b$ and obtain p from a t table or normal table, depending on the magnitude of d.f. and taking into account whether the test is one-tailed or two-tailed.

For example, suppose that $b = 1.7$, $s_b = .4$, and d.f. $= n - 2 = 998$. Suppose also that the null hypothesis is H_0: $\beta = 1$ (a hypothesis that arises in certain applications in economics) and that the alternative hypothesis is H_1: $\beta \neq 1$, so that the test is two-tailed. We have that $t = (b - \beta_0)/s_b = (1.7 - 1)/.4 = 1.75$. Because d.f. $\gg 120$ (the symbol "$\gg$" means "is large compared with"), we can assume that t is normally distributed and use the table of normal probabilities in Table B.2 (Appendix B). For $Z = 1.75$, the table indicates a one-tailed p of .0401, which we must double to arrive at a two-tailed p of .0802. p values are usually rounded to two decimal places, so we would write this as $p = .08$.

Because p values are probabilities, they range between zero and one. A low p value is a number near zero, and a high p value is a number near 1. *A low p value indicates a high observed level of statistical significance, and a high p value indicates a low observed level of statistical significance.*

1.5.5. Importance of a Good Spread of Values of the Predictor Variable

Other things being equal, a larger spread of values of the predictor variable means a higher likelihood that the regression coefficient, b, will be statistically significant at a specified level.

The importance of a good spread of values of the predictor variable is illustrated graphically in Figure 1.4. In the upper graph the variation in X is small, and in the lower graph it is large. The values of e_i for the three points, corresponding to deviations from the solid line,

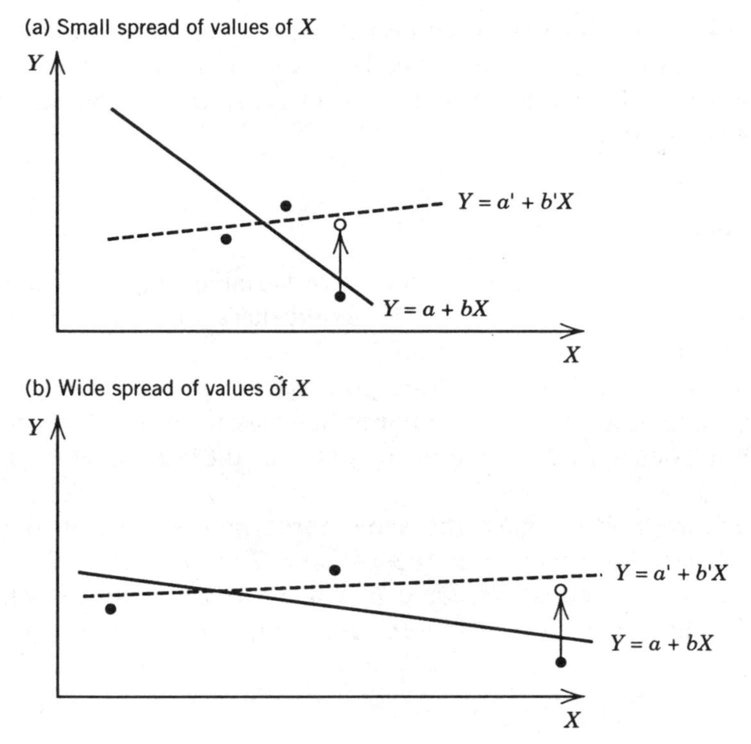

Figure 1.4. The effect of the spread of values of X on the statistical stability of the regression coefficient, b. **Note:** $Y = a' + b'X$ is the new line that is fitted when the rightmost point is moved upward as indicated by the arrow. The magnitude of the upward movement is the same in both graphs. The upward movement may be thought of as resulting from the drawing of another sample from the underlying population. The effect of the upward movement on the slope of the line is greater when the spread of X is smaller. The Y coordinates of the points are the same in the two graphs.

are the same in either graph. In each graph, what is the effect on b of the same upward change in Y for the rightmost point (due, say, to drawing another sample)? When the points are clustered along the X dimension, as in the upper graph, the effect on b is large, but when the points are spread out, the effect on b is small.

Figure 1.4 provides an intuitive understanding of why tight clustering on the X dimension results in a statistically unstable estimate of b, in the sense of high sampling variability of b. This instability is reflected in a high standard error, a low t value, and a high p value for b.

Unfortunately, in observational research there is nothing we can do to expand the spread of the predictor variable. If the underlying population has a narrow spread on this variable, our sample will tend to have a narrow spread, too. The best we can do, if we have advance knowledge of the narrow spread, is to increase the sample size by way of compensation.

1.5.6. Beware of Outliers!

Although a good spread in the predictor variable, X, is desirable from the point of view of statistical significance, one must beware of outliers. The problem is illustrated in Figure 1.5.

In Figure 1.5a there are three points that are tightly clustered on the X dimension. The least-squares line has negative slope b, but b probably does not differ significantly (say at the 5 percent level) from zero.

In Figure 1.5b we have the same three points, plus an outlier (a point that is far away from the others). *The outlier exerts a strong influence on the estimated slope*, b, which is now positive. The estimated value of b is now quite different, and b may now differ

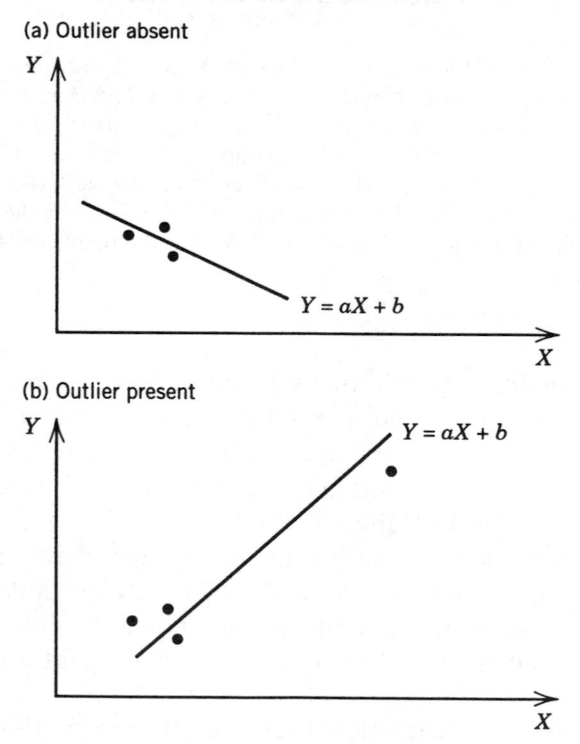

Figure 1.5. Illustration of the effect of an outlier on the fitted least-squares line.

significantly from zero. The outlier may also exert a strong influence on the constant term, a.

How does one deal with outliers? First of all, it is necessary to detect them. To do this, it is a good idea to have the computer routinely print out a *scatterplot* (a plot of points like that shown in Figure 1.5a or 1.5b). A scatterplot is also useful for checking visually whether the assumption of a straight-line relationship is reasonable.

If the sample is too large for a scatterplot to be feasible, another possibility is to have the computer tabulate frequency distributions. To detect outliers on the X dimension, divide the X dimension into suitable intervals and tabulate the number of observations in each interval. Then do the same for Y.

Most computer packages for regression routinely print out minimum and maximum value for each variable. Perusal of these values may also be useful for detecting outliers.

These preliminary explorations of the data should be done routinely, before running regressions. The explorations are easily accomplished with the aid of the scatterplot and frequency count options in such statistical computing packages as SPSS, SAS, BMDP, and LIMDEP.

Outliers are often due to data errors. For example, it is easy to forget to exclude unknowns, which are often coded "99." If the variable in question is, say, number of children ever born, 99 will be an outlier.

Once a genuine outlier is identified, the regression can be run with and without the outlier in order to assess its impact on the estimate of β. This impact will tend to be smaller the larger the sample size.

1.5.7. Beware of Selection on the Response Variable!

Often the sample is not perfectly representative of the underlying population. Certain characteristics may be underrepresented or overrepresented in the data. For example, in a telephone survey, people below the poverty line may be underrepresented, because they are less likely to have a telephone than people above the poverty line. In this case the sample suffers from selection on income. Selection problems also occur frequently in the analysis of retrospective survey data. An example is marriage history data; here the problem is that the sample members were younger in the past, so that there is selection on age as one goes back in time. For instance, if the original sample was of women aged 15–49, then 30 years ago these same women were aged 0–19, at which time all their marriages to date must have occurred before age 20. This age selectivity presents problems if one wishes to use the retrospective data to estimate the trend in mean age at marriage during the past 30 years.

How does selection affect the application of the regression model to the sample data? *As long as the assumption of linearity is met, selection on the predictor variable is not a problem.* The expected value of Y at each X will be unaffected by the selection, and the regression line will be unchanged. Even if some X's are completely selected out of the sample, the expected values of Y at the remaining X's will be unchanged, and the line will therefore be unchanged.

The situation can change radically if there is selection on the response variable Y. Then the mean of Y may be thrown off at some or perhaps all of the X's. The slope b, the intercept a, or both may be seriously biased because of the selection.

This means, for example, that one would not want to treat migration as the response variable in a retrospective survey of past moves based on a sample that oversampled migrants.

What if the sample is a stratified sample? To the extent that some strata are intentionally oversampled, stratification introduces an element of selection on whatever variables are used to define the strata. We may call such a variable a *stratification variable*. As long as the stratification variable is used as a predictor variable, there will be no problem. *But the stratification variable should not be used as the response variable.*

The observation that selectivity on Y is a problem, whereas selectivity on X is not, is consistent with the sampling assumptions of the bivariate regression model, discussed earlier.

1.5.8. Presentation of Results

Sometimes the coefficient b is simply marked by one, two, or three asterisks to indicate that b differs significantly from zero at the 5, 1, or 0.1 percent level of significance. For example, if $b = -0.3$, the tabulated value of b might appear as -0.3^* in a table of results. A footnote to the table might explain that *** means $p \leq .001$, ** means $.001 < p \leq .01$, and * means $.01 < p \leq .05$.

Alternatively the standard error, t value, or p value of b might be indicated in parentheses. for example, the results of fitting the regression might be presented as

$$Y = 4.2 - \underset{(0.1)}{0.3X} \tag{1.23}$$

where 0.1 is the standard error of b, which in this sample has the value of -0.3.

Each of these ways of presenting the results can be found in the literature. Although there is no agreed-upon standard mode of presentation, in this book we shall usually present standard errors in parentheses and mark coefficients with an asterisk in the event that $p < .05$. An advantage of presenting the standard error s_b is that it is easy to calculate in one's head an approximate 95 percent confidence interval as $b \pm 2s_b$ to get a feeling for the precision of b as an estimate of β.

In (1.23), the confidence interval obtained in this way is $(-.5, -.1)$. Because 0 does not fall in this interval, one can immediately see that the coefficient differs significantly from 0 at the 5 percent level, assuming a two-tailed test. One can also calculate in one's head that $b/s_b = -.3/.1 = -3$, which is considerably less than -1.96, providing another quick indicator (if d.f. exceeds 120 or so) that b differs significantly from 0 at the 5 percent level.

Presenting observed p values (instead of standard errors) in parentheses is a more concise but less informative way of indicating statistical significance that we shall also use on occasion.

In Sections 1.5.1–1.5.8 we have considered how one can generalize from b to β. What about generalizing from a to α? The approach is the same, except that we must use s_a instead of s_b when calculating a t value. But generalizing from a to α is less interesting than generalizing from b to β, because we are usually interested in causal analysis, and, if we are using the bivariate regression model to portray the effect of X on Y, b is the measure of effect.

1.6. GOODNESS OF FIT

How closely do the points fit the regression line? We consider two measures: the standard error of the estimate, s, and the correlation coefficient, r.

1.6.1. Standard Error of the Estimate, s

One measure of goodness of fit is σ, the *standard deviation* (the square root of the variance, σ^2) of the errors, ε_i, in the underlying population:

$$\sigma = \sqrt{\sigma^2} = \sqrt{\frac{\Sigma \varepsilon_i^2}{N}} \qquad (1.24)$$

where N is the size of the underlying population and the summation ranges over the entire population. (See the homoscedasticity assumption in Section 1.5.1.) $(\Sigma\varepsilon_i^2)/N$ is the average squared error, and the square root of this quantity may be interpreted as a kind of average error.

When the calculation is done from sample data, it can be shown that

$$s^2 = \frac{\Sigma e_i^2}{n - 2} \tag{1.25}$$

is an unbiased estimator of σ^2 (Theil, 1971). In (1.25), n denotes the size of the sample, and the summation ranges over the sample. The square root of s^2, namely s, is called the *standard error of the estimate*, and it is sometimes denoted alternatively as s_e, SEE, or root MSE (mean squared error). An advantage of s as a measure of goodness of fit is that it has a reasonably clear meaning as a kind of "average" absolute size of the residuals.

1.6.2. Coefficient of Determination, r^2, and Correlation Coefficient, r

Consider the following decomposition of $Y_i - \bar{Y}$, as illustrated in Figure 1.6:

$$\boxed{Y_i - \bar{Y} = (Y_i - \hat{Y}_i) + (\hat{Y}_i - \bar{Y})} \tag{1.26}$$

This formula expresses the deviation of Y_i from $\bar{Y}$ as the sum of an *unexplained component* (also called a *residual component*), $e_i = Y_i - \hat{Y}_i$, and an *explained component*, $\hat{Y}_i - \bar{Y}$. By "explained" we mean that $\hat{Y}_i - \bar{Y}$ is that part of $Y_i - \bar{Y}$ which is accounted for by the regression line.

Squaring both sides of (1.26) and summing over all members of the sample, we obtain

$$\Sigma(Y_i - \bar{Y})^2 = \Sigma(Y_i - \hat{Y}_i)^2 + \Sigma(\hat{Y}_i - \bar{Y})^2 + 2\Sigma(Y_i - \hat{Y}_i)(\hat{Y}_i - \bar{Y}) \tag{1.27}$$

It can be shown that the last term on the right side is zero. After rearrangement of remaining terms on the right side, (1.27) can be

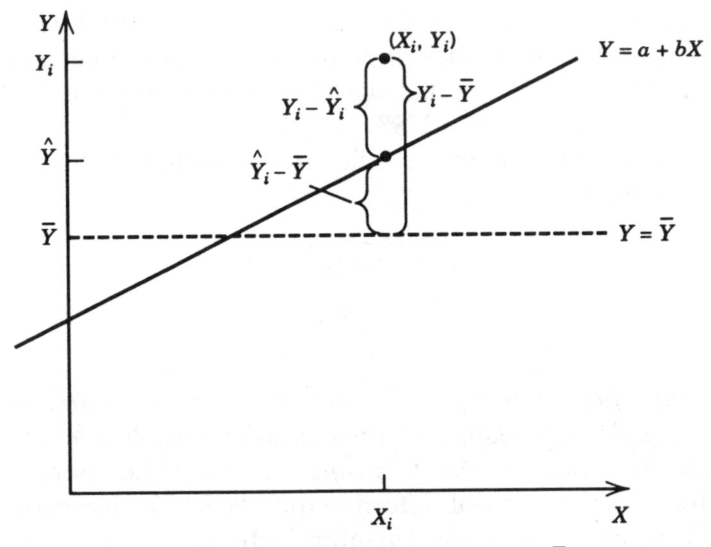

Figure 1.6. Decomposition of $Y_i - \bar{Y}$.

rewritten as

$$\Sigma\left(Y_i - \bar{Y}\right)^2 = \Sigma\left(\hat{Y}_i - \bar{Y}\right)^2 + \Sigma\left(Y_i - \hat{Y}_i\right)^2 \tag{1.28}$$

The terms in (1.28) have the following names:

$\Sigma(Y_i - \bar{Y})^2$: *total sum of squares*, or *TSS*; also called *total variation*

$\Sigma(\hat{Y}_i - \bar{Y})^2$: *explained sum of squares*, or *ESS*; also called *explained variation*

$\Sigma(Y_i - \hat{Y}_i)^2$: *unexplained sum of squares*, or *USS*; also called *unexplained variation*

Sometimes ESS is called the *regression sum of squares* or *model sum of squares*, and USS the *residual sum of squares* or *error sum of squares*.

With these definitions, (1.28) can be rewritten

$$\text{TSS} = \text{ESS} + \text{USS} \tag{1.29}$$

or, in words, *total variation equals explained variation plus unexplained variation*.

If all the points fall on the fitted line, TSS = ESS and USS = 0. At the other extreme, if the scatter of points resembles a circular cloud, it will be found that $b = 0$, so that the fitted line reduces to $Y = \overline{Y}$. In this case ESS = 0 and TSS = USS.

These observations lead to the following definition of the *coefficient of determination*, r^2:

$$r^2 \equiv \frac{\text{ESS}}{\text{TSS}} \tag{1.30}$$

The coefficient of determination, r^2, *measures the proportion of the total sum of squares that is explained by the regression line. It is a measure of how closely the points fit the least-squares line.* [The more general notation for the coefficient of determination is R^2, as discussed in the chapter on multiple regression (Chapter 2). In the bivariate case, both R^2 and r^2 are used.]

The *correlation coefficient*, r, may be defined as the square root of the coefficient of determination. r is assigned a positive sign if b is positive, and it is given a negative sign if b is negative.

If $b > 0$ and all the points fall on the fitted line, ESS = TSS and $r = 1$. If $b < 0$ and all the points fall on the fitted line, again ESS = TSS and $r = -1$. If there is no relationship between X and Y ($b = 0$), ESS = 0 and $r = 0$. *Thus the range of* r *is* $-1 \leq r \leq 1$. Because r is simply the square root of r^2, *r also measures how closely the points fit the least-squares line.*

If $r = 1$ or $r = -1$, we speak of X and Y as being *perfectly correlated*. If $r = 0$, we speak of X and Y as being *uncorrelated*. If X and Y are positively correlated ($r > 0$), Y tends to increase as X increases. If X and Y are negatively correlated ($r < 0$), Y tends to decrease as X increases.

Note that *r is unitless*. This is so because the numerator and denominator of (1.30) have the same units (i.e., squared units of Y), which cancel.

It can be shown that r can also be expressed as

$$r = \frac{\Sigma(X_i - \overline{X})(Y_i - \overline{Y})}{\sqrt{\Sigma(X_i - \overline{X})^2 \Sigma(Y_i - \overline{Y})^2}} \tag{1.31}$$

and in fact this equation is usually used as the starting definition of the correlation coefficient between X and Y. Equivalently,

$$\boxed{r = \frac{s_{XY}}{s_X s_Y}} \tag{1.32}$$

where

$$s_{XY} \equiv \frac{\Sigma(X - \bar{X})(Y - \bar{Y})}{n - 1} \tag{1.33}$$

$$s_X \equiv \sqrt{\frac{\Sigma(X - \bar{X})^2}{n - 1}} \tag{1.34}$$

$$s_Y \equiv \sqrt{\frac{\Sigma(Y - \bar{Y})^2}{n - 1}} \tag{1.35}$$

s_{XY} is the *sample covariance* of X and Y, s_X is the *sample standard deviation of* X, and s_Y is the *sample standard deviation of* Y. The formulae for σ_{XY}, σ_X, and σ_Y in the underlying population are the same as those given in (1.33)–(1.35), except that the summation is over all members of the population instead of the sample, and the divisor is N, the total population size, instead of $n - 1$.

It can also be shown that

$$\boxed{r = b\frac{s_X}{s_Y}} \tag{1.36}$$

This equation demonstrates that when $b = 0$, *r must also be zero, and vice versa*. A good way to remember that the multiplier is s_X/s_Y and not s_Y/s_X is to remember that r is unitless. Because the units of b are Y units/X units (b is the change in Y per unit change in X), the multiplier of b must be s_X/s_Y so that units cancel in the product, leaving a unitless number.

Although r and b are related as in (1.36), they have quite different meanings that are sometimes confused. *The correlation coefficient r is a measure of goodness of fit. In contrast, in the context of causal modeling, b is a measure of effect. These are two very different concepts.*

1.7. FURTHER READING

For those without much mathematics background, a clear and easy-to-read development of basic probability theory is found in Wonnacott and Wonnacott (1984), *Introductory Statistics for Business and Economics*. A clear and easy-to-read development of bivariate linear regression and correlation is found in Wonnacott and Wonnacott (1979), *Econometrics*, and in Hanushek and Jackson (1977), *Statistical Methods for Social Scientists*.

For those familiar with matrices and vectors, a somewhat more advanced treatment of regression is found in Fox (1984), *Linear Statistical Models and Related Methods*, and a considerably more advanced treatment in Theil (1971), *Principles of Econometrics*. More rigorous treatments may be found in Mood, Graybill, and Bose (1974), *Introduction to the Theory of Statistics*, and Stuart and Ord (1987 and 1991, in two volumes), *Kendall's Advanced Theory of Statistics*.

An instructive discussion of the issue of causality is found in a survey article by Holland (1986).

2

MULTIPLE REGRESSION

A problem with the bivariate linear regression model described in Chapter 1 is that, in the context of causal analysis, a causal model with a single predictor variable is rarely adequate for analyzing real-world behavior. Multiple regression solves this problem by extending the bivariate linear regression model to include more than one predictor variable.

The general form of the multiple regression model is

$$\boxed{Y_i = \beta_0 + \beta_1 X_{1i} + \beta_2 X_{2i} + \cdots + \beta_k X_{ki} + \varepsilon_i} \tag{2.1a}$$

where k denotes the number of predictor variables and i denotes the ith member of the population. The corresponding estimated model, pertaining to a particular sample from the population, is

$$\boxed{Y_i = b_0 + b_1 X_{1i} + b_2 X_{2i} + \cdots + b_k X_{ki} + e_i} \tag{2.1b}$$

For (2.1a) and (2.1b) to serve as a statistical model, we need to invoke again the assumptions of *linearity*, *homoscedasticity*, and *independence*. "Linearity" now means that the mean of the disturbances is zero at each combination of the predictor variables; that is, $E(\varepsilon) = 0$ at every point $(X_1, X_2, \ldots, X_k)$. "Homoscedasticity" means that the variance of ε is the same for each combination of the predictor variables; that is, $\text{Var}(\varepsilon) = \sigma^2$ at every point $(X_1, X_2, \ldots, X_k)$. "Independence" again means that the ε_i are statistically independent. Sampling assumptions are the same as those for bivariate linear regression, with random sampling required for the Y's but not the X's.

Although in this book Y is usually referred to as the *response variable* and $X_1, \ldots, X_k$ as the predictor variables, Y is also referred to as the *dependent variable* and $X_1, \ldots, X_k$ as the *independent variables. In this context, "independent" does not mean "statistically independent." The variables* $X_1, \ldots, X_k$ *may be intercorrelated and therefore not statistically independent.*

2.1. THE PROBLEM OF BIAS IN BIVARIATE LINEAR REGRESSION

An advantage of (2.1a) and (2.1b), compared with the simple bivariate linear model

$$Y_i = \beta_0 + \beta_1 X_{1i} + \varepsilon_i \qquad \text{(population)} \tag{2.2a}$$

$$Y_i = b_0 + b_1 X_{1i} + e_i \qquad \text{(sample)} \tag{2.2b}$$

is that the inclusion of more predictor variables tends to reduce the size of the disturbances ε_i and the residuals e_i. *A more important advantage, however, is that the inclusion of more predictor variables tends to reduce bias.*

As an example of how bias can occur, consider the model

$$Y = \beta_0 + \beta_1 X_1 + \beta_2 X_2 + \varepsilon \qquad \text{(population)} \tag{2.3a}$$

$$Y = b_0 + b_1 X_1 + b_2 X_2 + e \qquad \text{(sample)} \tag{2.3b}$$

where

$$Y = \text{number of children ever born (fertility)}$$

$$X_1 = \text{income}$$

$$X_2 = \text{number of completed years of education}$$

(For simplicity, subscript i, denoting the ith member of the population or sample, is now dropped.) For purposes of illustration, let us suppose that the underlying assumptions of the model are met, and that the model adequately portrays the effects of income and education on fertility. Let us also suppose that X_1 and X_2 are positively correlated (the higher income is, the higher education is, and vice versa), β_1 is positive (income has a positive effect on fertility), and β_2 is negative (education has a negative effect on fertility).

If we erroneously omit, say, X_2 from (2.3a) and (2.3b), the model becomes bivariate as in (2.2a) and (2.2b). *When we fit this misspecified model to sample data, we will find that b_1 is less positive than before, or even negative.* The problem is that part of the apparent effect of X_1 is actually due to the correlated but omitted variable X_2. In other words, income has a true positive effect when education is included in the model, but a less positive or even negative effect on fertility when education is erroneously omitted from the model. When education is not in the model, the apparent effect of income includes part of the effect of education (because education and income are correlated), and this effect of education, which is negative, lessens or reverses the true positive effect of income on fertility. *In this example, b_1 is a biased estimator of β_1 when education is not in the model, and the bias is in the negative direction.*

There may be even more omitted variables that affect fertility, but as long as they are not correlated with the predictor variables X_1 and X_2, their omission does not bias the estimates of b_1 and b_2 in (2.3b).

2.2. MULTIPLE REGRESSION WITH TWO PREDICTOR VARIABLES

How do we fit the multiple regression model? The simplest case is multiple regression with two predictor variables. Let us consider the estimated model

$$\hat{Z} = a + bX + cY \tag{2.4}$$

where

Z = number of children ever born (fertility)

X = income

Y = education

Equation (2.4) is the same as (2.3b); we have simply changed the variable names to avoid writing subscripts, and we have substituted $\hat{Z} + e$ for Z.

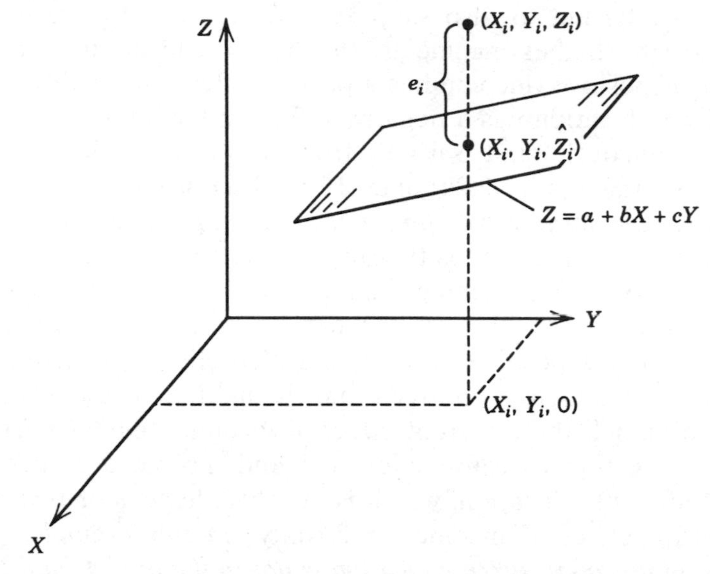

Figure 2.1. Multiple regression with two predictor variables: fitting a least-squares plane.

We can represent X, Y, and Z on coordinate axes, as in Figure 2.1. In three-dimensional space (3-space for short), (2.4) is the equation of a plane. For every point (X_i, Y_i) in the X–Y plane, the model in (2.4) generates an estimated value of Z, namely $\hat{Z}_i$.

How do we fit the plane through the scatter of observed points (X_i, Y_i, Z_i) in 3-space? Again we form the sum of squared residuals:

$$S = \sum e_i^2 = \sum \left(Z_i - \hat{Z}_i\right)^2 = \sum (Z_i - a - bX_i - cY_i)^2 \quad (2.5)$$

The best-fitting values of a, b, and c are those that minimize S. To find these values, we compute three partial derivatives of S with respect to a, b, and c, respectively. We set each partial derivative equal to zero, thereby obtaining three equations in the three unknowns a, b, and c. These normal equations can be solved for a, b, and c. Details are not shown. In practice, we let the computer do the work.

Suppose that we have fitted the model in (2.4), so that we have numerical values of a, b, and c. How do we interpret the coefficients b and c of X and Y?

We proceed as before by asking, what is the effect on Z of increasing X by one unit? Denoting the new value of Z by $\hat{Z}^*$, we obtain

$$\hat{Z}^* = a + b(X + 1) + cY = a + bX + cY + b = \hat{Z} + b \quad (2.6)$$

We see that b is the increase in Z per unit increase in X, holding Y constant. Note that the value of the fixed Y doesn't matter; the increase in Z per unit increase in X is always the same. By similar reasoning, c is the increase in Z per unit increase in Y, holding X constant.

Suppose that we have a theory that income (X) and education (Y) are causes of fertility (Z). *Then we interpret b as the estimated effect on fertility of one additional unit of income, holding education constant, and c as the estimated effect on fertility of one additional unit of education, holding income constant.*

These interpretations of b and c are very nice: We have found a statistical method for estimating the effect of one variable while *controlling* for other variables (i.e., holding them constant in a statistical sense), without doing a controlled experiment. Of course, these interpretations are valid only to the extent that both the theory and the assumptions underlying the model are valid.

Because the coefficient, b or c, of each predictor variable incorporates controls for the other predictor variable, these multiple regression coefficients are often referred to alternatively as *partial regression coefficients*.

The model in (2.4) is an *additive model* in at least two ways. First, a one-unit increase in, say, X generates an additive change of b units in Z, as shown in (2.6). The second sense of additivity pertains to variation of both explanatory variables simultaneously. Suppose we increase X and Y each by one unit simultaneously. Denoting the new value of Z by $\hat{Z}^*$, we have

$$\begin{aligned} \hat{Z}^* &= a + b(X + 1) + c(Y + 1) = a + bx + cY + b + c \\ &= \hat{Z} + (b + c) \end{aligned} \quad (2.7)$$

The joint effect is the sum of the separate effects.

By holding either X or Y constant, we can gain some insight about why the model $\hat{Z} = a + bX + cY$ in (2.4) is considered a linear model. If Y is held constant at Y_0, (2.4) becomes

$$\hat{Z} = (a + cY_0) + bX$$

which can be written

$$\hat{Z} = a^* + bX \tag{2.8}$$

where the constant term is now $a^* = a + cY_0$. Equation (2.8) represents a straight-line relationship between Z and X. Similarly, the relationship between Z and Y is a straight line when X is held constant. *In other words, when we look at the effect of one predictor variable on the response variable, holding the other predictor variables constant, the multiple regression model reduces to a straight line.* This is so no matter how many predictor variables are included in the model.

2.3. MULTIPLE REGRESSION WITH THREE OR MORE PREDICTOR VARIABLES

The approach is similar. Although a simple geometric representation similar to Figure 2.1 is no longer possible, the algebraic extension is straightforward. Fitting the model again involves minimizing the sum of squared residuals between estimated and observed values of the response variable. In general, if there are k predictor variables, one must set $k + 1$ partial derivatives equal to zero and solve $k + 1$ equations for $k + 1$ unknown regression parameters. In practice, we let the computer do the work. The interpretation of results is also similar.

2.4. DUMMY VARIABLES TO REPRESENT CATEGORICAL VARIABLES

Dummy variables (also called *dichotomous variables*, *binary variables*, or *contrast variables*) are used to represent categorical variables such as sex (male, female) and occupation (e.g., white collar, blue collar, farm). Dummy variables take on only two values, usually 0 and 1.

2.4.1. Categorical Variables with Two Categories

Categorical variables with two categories can be represented by a single dummy variable. The variable, urban–rural residence, is an example:

U: 1 if urban, 0 otherwise

The category with assigned value 0 is called the *reference category*. In the above example, rural is the reference category. However, we might just as well have defined urban as the reference category:

$$R: \quad 1 \text{ if rural, } 0 \text{ otherwise}$$

From the point of view of regression analysis, it makes no difference which category is chosen as the reference category. Suppose, for example, that we regress fertility, F, on U:

$$\hat{F} = a + bU \tag{2.9}$$

Because $U = 1 - R$, we can derive the regression of F on R directly from (2.9) as

$$\hat{F} = a + b(1 - R) = (a + b) - bR = a^* - bR \tag{2.10}$$

where $a^* = a + b$. We need not rerun the regression of F on R.

The coefficient of residence is b if residence is defined as U, and $-b$ if residence is defined as R. In (2.9), b is the effect on F of a one-unit increase in U. The only one-unit increase in U that is possible is from 0 (rural) to 1 (urban). Suppose that $b = -2$. Then (2.9) says that the effect of being urban, relative to rural, is to lower fertility by two children. Similarly, (2.10) says that the effect of being rural, relative to urban, is to raise fertility by two children. Clearly these two statements are equivalent. The interpretation of the intercepts in (2.9) and (2.10) can similarly be shown to be equivalent. In (2.9) the intercept a is the fertility of the reference category, which is rural. In (2.10) the intercept $a^* = a + b$ is the fertility of the reference category, which is urban. The two intercepts are equivalent in the sense that in either case, a is the fertility of rural and $a + b$ is the fertility of urban. *It is in these senses of equivalence that the choice of reference category is arbitrary*.

In defining dummy variables in general, why assign values of 0 and 1? Why not assign 1 and 2? Or 500 and 1000? We could do this, but there are two main reasons why 0 and 1 are preferred. To illustrate them, let us return to our hypothetical example in (2.9).

The first reason for using 0 and 1 has to do with the interpretation of a and b. Suppose residence is rural in (2.9), so that $U = 0$. Then $\hat{F} = a$. *a is the fertility of the reference category*. Suppose residence is urban, so that $U = 1$. Then $\hat{F} = a + b$. It follows that *b is the difference in fertility between urban and rural*. These are simple and convenient interpretations of a and b that are possible because we have assigned numerical values of 0 and 1. We would lose one or both of these interpretations if we chose 1 and 2 or 500 and 1000.

A second important reason for choosing 0 and 1 has to do with the interpretation of $\overline{U}$, the mean of U. We compute $\overline{U}$ by adding the values of U for all members of the sample and dividing by the number of members. If U is assigned values of 0 and 1 for rural and urban, $\overline{U}$ can be interpreted simply as the proportion of the sample that is urban. If the values of U were chosen as 1 and 2 or 500 and 1000, $\overline{U}$ would not have this convenient interpretation. *In sum, dummy variables are assigned values of* 0 *and* 1 *instead of other values for reasons of analytical convenience.*

2.4.2. Categorical Variables with More Than Two Categories

Dummy variables can also be used to represent categorical variables with more than two categories. Again a simple example will make the discussion more concrete and understandable.

Let us define an occupational variable with three categories: white collar, blue collar, and farm. We use two dummy variables to represent these three categories:

W: 1 if white collar, 0 otherwise

B: 1 if blue collar, 0 otherwise

Then the three categories of occupation are represented as

White collar: $W = 1,\ B = 0$

Blue collar: $W = 0,\ B = 1$

Farm: $W = 0,\ B = 0$

The category for which both dummies are zero, which in this case is the farm category, is the reference category.

In the example of occupation, it takes two dummy variables to represent three occupational categories. It is easy to see that, in general, *it takes $j - 1$ dummy variables to represent a variable with j categories.*

How do we regress fertility on occupation? Denoting fertility by F (not to be confused with farm), we have

$$\hat{F} = a + bW + cB \tag{2.11}$$

This equation is fitted in the usual way.

How do we interpret a, b, and c? Again for purposes of illustration, let us suppose that (2.11) adequately portrays the effect of occupation on fertility, even though (2.11) is clearly oversimplified. Substituting appropriate combinations of values of W and B into (2.11), we obtain the following values of F:

$$\begin{array}{lll} \text{White collar} & (W = 1, B = 0): & \hat{F} = a + b \\ \text{Blue collar} & (W = 0, B = 1): & \hat{F} = a + c \\ \text{Farm} & (W = 0, B = 0): & \hat{F} = a \end{array} \tag{2.12}$$

The intercept a is the fertility of the reference category, b is the fertility difference between white collar and the reference category, and c is the fertility difference between blue collar and the reference category. *In other words, b is the effect on F of being white collar, relative to the reference category, and c is the effect on F of being blue collar, relative to the reference category*. The extension to categorical variables with more than three categories is obvious.

What happens if we redefine the reference category to be blue collar? Denoting G as 1 if occupation is farm and 0 otherwise, we can then rewrite equation (2.11) as

$$\hat{F} = d + eW + fG \tag{2.13}$$

The intercept d is the value of F for the reference category, which is now blue collar. From (2.12), $d = a + c$. The coefficient e in (2.13) is the effect on F of being white collar, relative to blue collar. From (2.12), $e = (a + b) - (a + c) = b - c$. The coefficient f in (2.13) is the effect on F of being farm, relative to blue collar. From (2.12), $f = a - (a + c) = -c$. *It is not necessary to rerun the regression in* (2.13); *we can calculate d, e, and f from a, b, and c in* (2.11).

In (2.11) and (2.13), the effect of being white collar, relative to blue collar, is the same $(b - c)$, regardless of whether the reference category is farm or blue collar. Likewise the effect on F of being farm, relative to blue collar, is the same ($-c$), regardless of whether the reference category is farm or blue collar. *In general, estimated effects are not affected by the choice of reference category, and it is in this sense that the choice of reference category is arbitrary.*

2.5. MULTICOLLINEARITY

The concept of multicollinearity pertains to relationships among the predictor variables. *Two predictor variables are multicollinear if they are highly correlated. Three or more predictor variables are multicollinear if at least one of them can be expressed as a close-fitting linear function of the others.* For example, three predictor variables U, V, and W are multicollinear if the sample points closely fit the plane $U = a + bV + cW$.

Let us consider again the estimated model

$$\hat{Z} = a + bX + cY \qquad \text{(2.4 repeated)}$$

If X and Y are perfectly correlated, then their relationship conforms exactly to a straight line, $Y = m + nX$, and they are perfectly collinear. In this case, Z is defined only for those points in the X–Y plane that fall on the line $Y = m + nX$, as shown in Figure 2.2. Then the least-squares fit of (2.4) describes a line (rather than a plane) in 3-space. In Figure 2.2 this line is denoted by L. The vertical projection of L on the X–Y plane is the line $Y = m + nX$.

The reason why multicollinearity is a problem is immediately clear from the figure. *The problem is that any number of planes with varying degree of tilt can be fitted through the line L.* In this situation, the normal equations for a, b, and c are not uniquely solvable, and one receives the computer message, "singular matrix."

If X and Y are highly but not perfectly correlated, the equations are solvable, but the estimates of a, b, and c are imprecise in that they have large standard errors. The reason is that the points (X_i, Y_i) in the X–Y plane are tightly clustered about the fitted line $\hat{Y} = m + nX$, so that sampling variability is likely to have a large effect on the tilt of the fitted plane. In other words, the estimates of a, b, and c are statistically unstable. We have a high degree of multicollinearity, but not perfect multicollinearity.

The analogy in bivariate linear regression to multicollinearity is clustering of the X_i about a single value. If the clustering is perfect, so that $X_i = X_0$ for all i, then the only constraint on the fitted line is

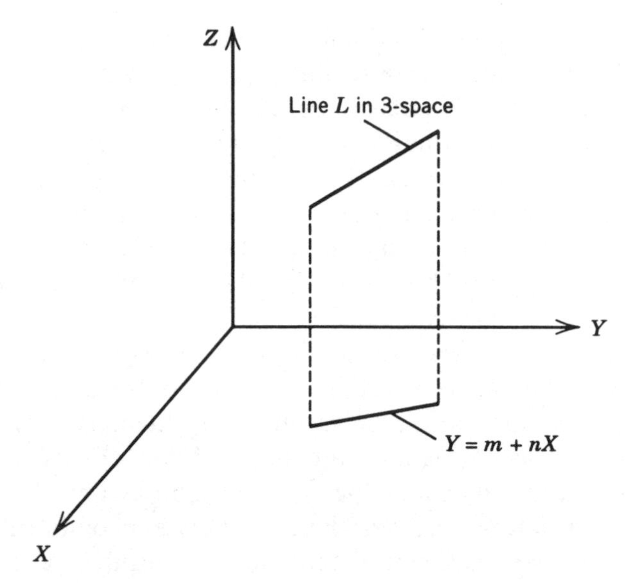

Figure 2.2. Illustration of multicollinearity. **Note:** All the values of (X_i, Y_i) fall on the line $Y = m + nX$. The problem is that any number of planes of varying tilt can be fitted through the line L.

that it pass through the mean value of Y at X_0. Because there is no spread in X, any line with any slope will do, as long as it passes through the point $(X_0, \bar{Y})$. If the X_i values are tightly but not perfectly clustered around X_0, the estimate of the slope of the fitted line is strongly affected by sampling variability. The estimate of the slope therefore has a high standard error and is statistically unstable. The other side of the coin, as discussed in Chapter 1, Section 1.5.5, is that a good spread of values of the predictor variable tends to increase the statistical precision of b (in the sense of a smaller value of s_b, which means a narrower confidence interval).

In sum, the problem of multicollinearity is one of statistical instability of coefficients due to inadequate spread. In bivariate linear regression, we need a good spread in the predictor variable. In multiple regression with two predictor variables, we need a good spread around the lines $\hat{Y} = m + nX$ and $\hat{X} = u + vY$. In multiple regression with more than two predictor variables, we need a good spread about each of the possible multiple regressions that can be fitted among the predictor variables.

One should try to specify multiple regression models so that the predictor variables arc not highly multicollinear. IIowever, in most real-life observational research (as opposed to experimental research, where treatments can be randomized), a certain amount of multicollinearity is inevitable, because most of the predictor variables in which we are interested (such as income and education) are correlated to some extent. As a rough rule of thumb, when two predictor variables are correlated but both are relevant to explanation from a theoretical point of view, one should not eliminate one of the variables to reduce multicollinearity, unless the correlations are higher in absolute magnitude than about 0.8.[1] In other words, we must live with a certain amount of multicollinearity.

Most computer programs for multiple regression can print out a *correlation matrix* as an optional part of the output. This is a table with the response and predictor variables listed both down the side and across the top as row labels and column labels. Entries within the table are correlations between pairs of variables. This table can be scanned for correlations greater than 0.8 between pairs of predictor variables. It must be emphasized that the 0.8 cutoff is a rough guide that may sometimes be inappropriate.

2.6. INTERACTION

Let us consider again a model with fertility (F) as the response variable and income (I) and education (E) as predictor variables. Previously we used Z, X, and Y to denote these variables. We now shift to F, I, and E, because the substantive meanings of these variable names are easier to remember.

What does it mean to say that there is interaction between two predictor variables? *Interaction between I and E means that the effect of I on F depends on the level of E, or that the effect of E on F depends on the level of I.* For example, there are situations where one expects the effect of education on fertility to be larger the higher the income. In general, theory will tell us whether to include an interaction term in the model.

2.6.1. Model Specification

The usual way of specifying interaction between I and E is to add a multiplicative IE term to the model:

$$\hat{F} = a + bI + cE + dIE \tag{2.14}$$

[1]Ideally one should examine more than just pairwise relationships among the predictor variables, but in practice the pairwise correlations usually tell most of the story.

Although this is the usual specification of interaction, it is not the only possible one. We could, for example, add an IE^2 term or a $\log(IE)$ term instead of an IE term. We choose IE because we have no reason to be more complicated than this. *We don't know the precise mathematical form of the interaction, so we choose as simple a form as possible.* This approach is in keeping with a general goal of modeling, which is to simplify reality down to its essentials so that we may more easily comprehend it.

In fitting the model in (2.14), we consider IE as if it were a third predictor variable, which we can denote by W. Thus $W = IE$. Then we fit the model

$$\hat{F} = a + bI + cE + dW \tag{2.15}$$

in the usual way.

What is the effect on F of increasing I by one unit? Denoting the new value of F by $\hat{F}^*$, we have from (2.14) that

$$\begin{aligned}\hat{F}^* &= a + b(I + 1) + cE + d(I + 1)E \\ &= a + bI + cE + dIE + b + dE \\ &= \hat{F} + (b + dE)\end{aligned} \tag{2.16}$$

The effect of a one-unit increase in I is to increase F by $b + dE$. Thus the effect on F of increasing I by one unit depends on the level of E, which is interaction by our earlier definition. Sometimes b is called the *main effect* and dE is called the *interaction effect*. Sometimes the coefficient d by itself is called the interaction effect. These differences in terminology can be confusing. One must not lose sight of the fact that the overall effect is $b + dE$.

Using calculus, we can derive the effect of I on F more simply by differentiating $\hat{F}$ with respect to I in equation (2.14). Again the result is $b + dE$.

We can similarly show that the effect on F of a one-unit change in E is $c + dI$. The symmetry of the two effects, $b + dE$ and $c + dI$, occurs because of the symmetry of the IE term itself. Had we used an IE^2 term instead of an IE term to model the interaction, the two effects would be asymmetric in mathematical form.

When the interaction term dIE is omitted, the model becomes

$$\hat{F} = a + bI + cE \tag{2.17}$$

The effect on F of a one-unit increase in I is simply b, as shown earlier. The effect is the same regardless of the level of E, so there is no interaction. Similarly the effect on F of a one-unit increase in E is c, regardless of the level of I, so again there is no interaction.

As shown earlier, (2.17) is an additive model, insofar as simultaneous one-unit increases in I and E generate an additive change in F of $b + c$, which is the sum of the effects on F of separate one-unit increases in I and E. But when there is an interaction term as in (2.14), separate one-unit increases in I and E generate changes in F of $b + dE$ and $c + dI$, which sum to $b + c + d(I + E)$. By way of comparison, we may derive the effect on F of simultaneous one-unit increases in I and E. Denoting the new value of F by $\hat{F}^*$, we have

$$\begin{aligned}\hat{F}^* &= a + b(I + 1) + c(E + 1) + d(I + 1)(E + 1) \\ &= a + bI + cE + dIE + b + c + d(I + E + 1) \\ &= \hat{F} + [b + c + d(I + E + 1)] \qquad (2.18)\end{aligned}$$

The simultaneous effect, $b + c + d(I + E + 1)$, differs from the sum of the separate effects, $b + c + d(I + E)$, so that effects are no longer additive. *Instead of an additive model, we now have an interactive model.*

Let us now view E as a control variable. Suppose $E = 2$. Then (2.14) becomes

$$\hat{F} = (a + 2c) + (b + 2d)I \qquad (2.19)$$

This is the equation of a line with intercept $a + 2c$ and slope $b + 2d$.

Suppose $E = 10$. Then (2.14) becomes

$$\hat{F} = (a + 10c) + (b + 10d)I \qquad (2.20)$$

Again we have a line, but the intercept and slope differ from those in (2.19). Suppose that the main effect of income on fertility is positive ($b > 0$), the main effect of education on fertility is negative ($c < 0$), and the effect of income on fertility decreases as education increases [$d < 0$, as is evident from (2.16)]. This situation is portrayed graphically in Figure 2.3.

Because $d < 0$, $b + 10d < b + 2d$. Thus the slope of the line corresponding to $E = 10$ is less than the slope of the line corresponding to $E = 2$. This agrees with our verbal specification that the effect of income on fertility decreases as education increases. Because $c < 0$, $a + 10c < a + 2c$. Thus the F intercept of the line corresponding to $E = 10$ is lower than the intercept of the line correspond-

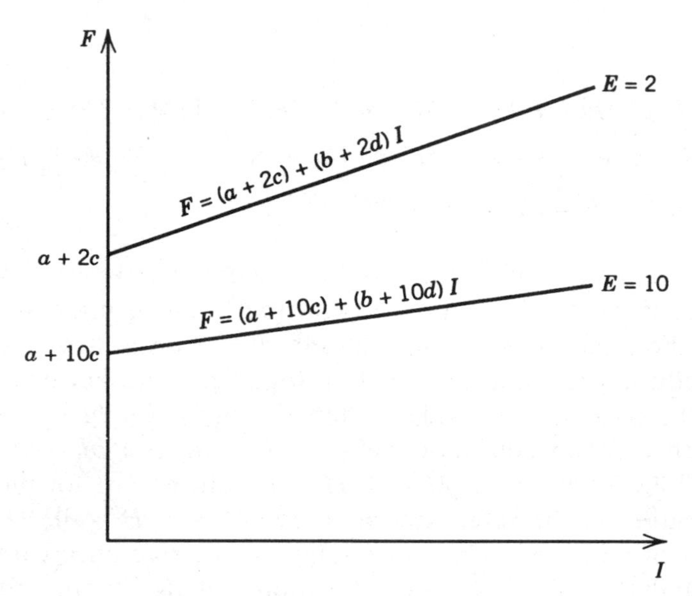

Figure 2.3. Graphical portrayal of interaction between I and E in the model $F = a + bI + cE + dIE$, for $E = 2$ and $E = 10$. **Note:** It is assumed that $a > 0$, $b > 0$, $c < 0$, and $d < 0$.

ing to $E = 2$. *We see that, with interaction specified as IE, controlling for E results in a simple linear relationship between F and I. This linear relationship changes as the level of the control variable changes.* Similarly, controlling for I results in a simple linear relationship between F and E, which changes according to the level of I.

2.6.2. More Complicated Interactions

Suppose that education is redefined as a categorical variable with three categories: medium (M), high (H), and low (reference category). Suppose also that education and income (I) interact in their effect on fertility. The estimated model is

$$\hat{F} = a + bI + cM + dH + eIM + fIH \tag{2.21}$$

The predictor variables now include multiplicative terms for all two-way combinations of I, on the one hand, and M or H, on the other.

What is the effect on F of increasing I by one unit, with education (M and H) held constant? Defining the new value of F by $\hat{F}^*$, we have

$$\begin{aligned}\hat{F}^* &= a + b(I + 1) + cM + dH + e(I + 1)M + f(I + 1)H \\ &= (a + bI + cM + dH + eIM + fIH) + b + eM + fH \\ &= \hat{F} + b + eM + fH \qquad (2.22)\end{aligned}$$

The effect of a one-unit increase in I is to increase F by $b + eM + fH$. Thus the effect of I on F depends on the levels of both M and H, which is the same as saying that the effect of I on F depends on the level of education, because M and H together represent education.

From (2.21) it is also evident that F equals $a + bI + c + eI$ for those with medium education ($M = 1$, $H = 0$), $a + bI + d + fI$ for those with high education ($M = 0$, $H = 1$), and $a + bI$ for those with low education in the reference category ($M = 0$, $H = 0$). Therefore the effect of medium education, relative to the reference category, is $c + eI$, and the effect of high education, relative to the reference category, is $d + fI$.

Suppose that income is also categorized into low, medium, and high, denoted by M' and H'. Then the model becomes

$$\hat{F} = a + bM + cH + dM' + eH' + fMM' + gMH' + hHM' + iHH' \qquad (2.23)$$

The model now includes $2 \times 2 = 4$ multiplicative terms for all two-way combinations of M or H, on the one hand, and M' or H', on the other. The derivation of formulae for effects is left to the reader as an exercise.

Less commonly, one can consider 3-way or even higher-order interactions. Suppose, for example, that F depends on income (I), education (E), and age (A), where all variables are quantitative. Suppose also that the effect of I on F depends on the combined levels of both E and A. A possible model of this 3-way interaction would be

$$\hat{F} = a + bI + cE + dA + eIE + fIA + gEA + hIEA \qquad (2.24)$$

If a 3-way interaction term is included in the model, the 2-way interaction terms that are embedded in the 3-way interaction term are usually also included, because 3-way interactions, if actually present, are usually accompanied by 2-way interactions. The logic for deriving formulae for effects is the same as before and is again left to the reader as an exercise.

2.6.3. Correlation without Interaction

The two concepts of correlation and interaction are sometimes confused with each other, but they mean very different things. In our earlier example of the effects of income and education on fertility, correlation between I and E means that I and E tend to be related as a straight line. Interaction between I and E means that the effect of I on F depends on the level of E, or that the effect of E on F depends on the level of I.

In the quite plausible model in (2.17), I and E are positively correlated, but there is no interaction term:

$$\hat{F} = a + bI + cE \qquad \text{(2.17 repeated)}$$

The effect of I on F does not depend on the level of E, and the effect of E on F does not depend on the level of I. Thus (2.17) illustrates that it is possible to have correlation without interaction.

2.6.4. Interaction without Correlation

It is also possible to have interaction without correlation. Consider the following model:

$$\hat{I} = a + bM + cU + dMU \qquad (2.25)$$

where

$$\begin{aligned} I &= \text{income} \\ M &= \text{sex (1 if male, 0 otherwise)} \\ U &= \text{residence (1 if urban, 0 otherwise)} \end{aligned}$$

Suppose that sex composition (percent male) is the same in urban and rural areas; that is, in urban areas, a person is no more likely to be male than in rural areas. Suppose also that there is more sex discrimination in wages in rural than in urban areas, so that the effect of sex on income is greater in rural than in urban areas.

In other words, the effect of M on I depends on the level of U, so that there is interaction between M and U. The simplest way to specify the interaction is by including a MU term in the model, as in (2.25). But sex and residence are not correlated, because sex composition does not vary by residence. In sum, M and U interact, but they are not correlated.

2.7. NONLINEARITIES

The effect of a predictor variable on the response variable is linear if the effect of a one-unit change in the predictor variable is the same, regardless of the level of the predictor variable. In the bivariate case, this means that the relationship between the response variable and the predictor variable can be portrayed as a straight line. *Conversely, the effect of a predictor variable is nonlinear if it varies according to the level of the predictor variable.*

If the relationship between, say, fertility (F) and education (E) is nonlinear, in the form of a curve instead of a straight line, what do we do? We consider two ways of capturing this kind of nonlinearity: first by adding a quadratic term in E, and second by using a dummy variable specification of education.

2.7.1. Quadratic Specification

Suppose the effect of education on fertility becomes increasingly negative at higher levels of education, as illustrated in Figure 2.4. The simplest way to introduce this kind of simple curvature into the model is to add a quadratic term (i.e., a squared term) to the model:

$$\hat{F} = a + bE + cE^2 \tag{2.26}$$

Alternatively we might add, say, an exponential or logarithmic term in

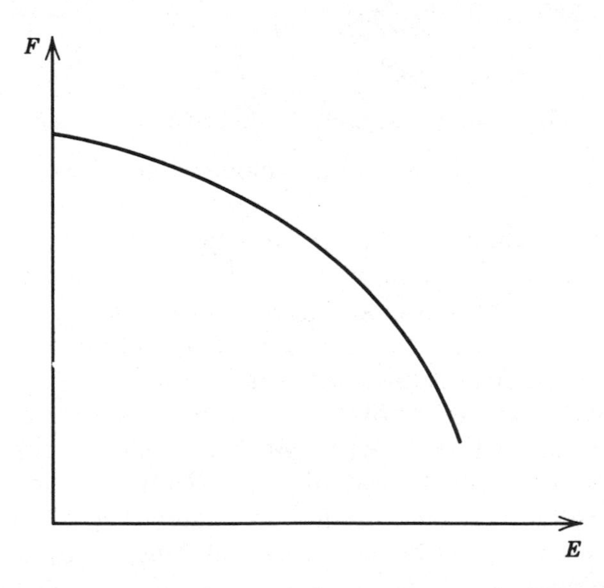

Figure 2.4. Illustration of a nonlinear relationship between fertility and education.

E. But there is usually no point in getting more complicated than a quadratic term, given the rather imprecise nature of our theory about how education affects fertility.

Equation (2.26) is the equation of a parabola that opens either upward or downward, depending on whether c is positive or negative. Figure 2.5 shows the general case, with the downward-opening parabola in Figure 2.5b positioned to model the relationship between F and E as depicted in Figure 2.4. By changing the values of a, b, and c in (2.26) appropriately, the parabola not only can be made to open upward or downward but also can be made wider or narrower or moved up, down, right, or left.

Figure 2.4, which shows the curvilinear relationship between F and E, does not look like a parabola. But this is not a problem, because *we do not use all of the parabola to fit the scatter of points, but only part of it*, as shown in Figure 2.5b. As long as the curvature we wish to fit approximates the form of a simple arc, we can usually find a piece of some parabola to fit it adequately.

How do we fit the model in (2.26)? We treat E^2 just as if it were another predictor variable, which we may again call W. Thus $W = E^2$.

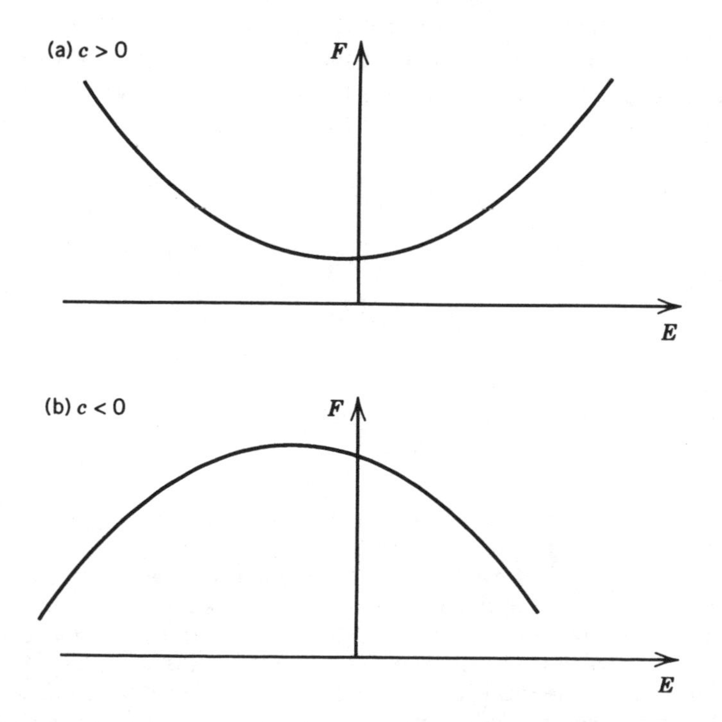

Figure 2.5. Graph of the parabola $F = a + bE + cE^2$. **Note:** The parabola in part b has been positioned to model the relationship between F and E in Figure 2.4.

Then (2.26) becomes

$$\hat{F} = a + bE + cW \tag{2.27}$$

which we fit in the usual way.[2]

But wait. Knowing E, we can predict E^2 exactly. Don't we have perfect multicollinearity?

No. We have perfect predictability of E^2 from E, but we don't have perfect multicollinearity, which would mean perfect correlation between E and E^2. *Predictability and correlation are not the same thing*. Perfect correlation means a perfect *linear* relationship between the variables. But the graph of E^2 against E is a parabola.

The geometry of (2.26) helps to explain why perfect predictability of E^2 from E is not a problem. Our earlier discussion of perfect multicollinearity included a graph (Figure 2.2) showing that the locus of predicted values of the response variable was a straight line in 3-space instead of a plane. Any number of planes of varying tilt could be fitted through the line L, so that the parameters of the plane, a, b, and c, could not be uniquely estimated. But when (2.26) is fitted, the locus of predicted values of F is a parabolic space curve. In this case, the curvature constrains the amount of tilt of the fitted plane that contains the curve, so that only one plane can pass through the curve.

What is the effect of a one-unit change in E on F in (2.26)? Denoting the new value of F by $\hat{F}^*$, we obtain

$$\begin{aligned} \hat{F}^* &= a + b(E + 1) + c(E + 1)^2 \\ &= a + bE + cE^2 + (b + 2cE + c) \\ &= \hat{F} + (b + 2cE + c) \end{aligned} \tag{2.28}$$

The effect of a one-unit increase in E is to change F by $b + 2cE + c$ units. *From this expression it is clear that the effect of E depends on the level of E, in accordance with our definition of a nonlinear effect.*

[2]Although equation (2.26) represents a parabola, from another perspective it is still a linear model, insofar as the model is still linear in the parameters to be estimated, namely a, b, and c. The term "linear" may refer to either linearity in the variables or linearity in the parameters. In the present instance, we have a model that is linear in the parameters but nonlinear in the variables. The interactive models considered earlier are also linear in the parameters but nonlinear in the variables. Models that are linear in the parameters can all be estimated by the same OLS methodology if the underlying assumptions of linearity, homoscedasticity, and independence are met. In this book we shall usually use the word "linear" to mean "linear in the variables."

Because by assumption, the effect of education on fertility is negative and becomes increasingly negative as education increases, it is evident in this particular example that both b and c must be negative.

If we take the derivative of $\hat{F}$ with respect to E, we get $d\hat{F}/dE = b + 2cE$, which is not quite the same as $b + 2cE + c$ in (2.28). The derivative is the *"instantaneous" rate of change* in F with respect to E; geometrically, it is the slope of the line tangent to the parabola at E. In contrast, $b + 2cE + c$ is the slope of the straight line (called a *chord*) connecting the two points on the parabola between E and $E + 1$. *We may consider the effect on F of a one-unit increase in E to be either $b + 2cE + c$ or $b + 2cE$, depending on whether we are referring to the "discrete" or the "instantaneous" effect.*

2.7.2. Dummy Variable Specification

Sometimes we use dummy variables to represent a quantitative predictor variable as a set of categories, in order to capture nonlinearities in the relationship between the predictor variable and the response variable.

Suppose again that the response variable is fertility (F) and the predictor variable is education (E). Suppose also, for illustrative purposes, that the relationship between fertility and education is such that fertility is constant up to a threshold level of education, after which fertility drops off at a constant rate with respect to education, as shown in Figure 2.6a.

One possible way to incorporate this kind of nonlinearity into our regression model is to add an E^2 term, which, as we have already seen, would generate a simple arc-like curvature that would roughly approximate the piecewise-linear "curve" in Figure 2.6a. (By "piecewise-linear" we mean a chain of straight line segments.) Alternatively we can use dummy variables.

To illustrate, let us first redefine education as follows:

M: 1 if medium education, 0 otherwise

H: 1 if high education, 0 otherwise

We may choose cutting points for low, medium, and high education as shown by tick marks on the horizontal axes of the graphs in Figure 2.6.

The estimated regression model is then

$$\hat{F} = a + bM + cH \qquad (2.29)$$

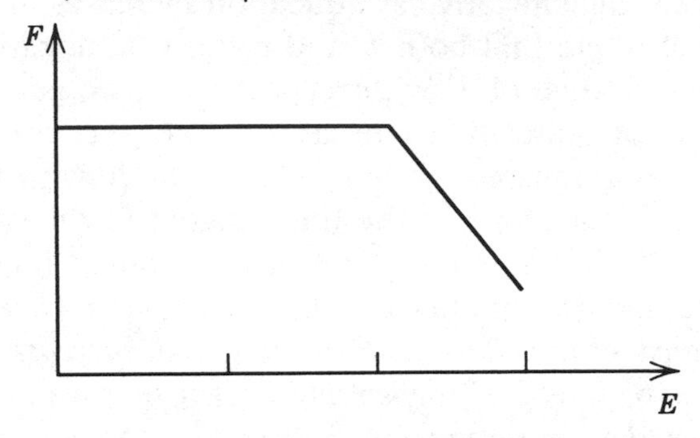

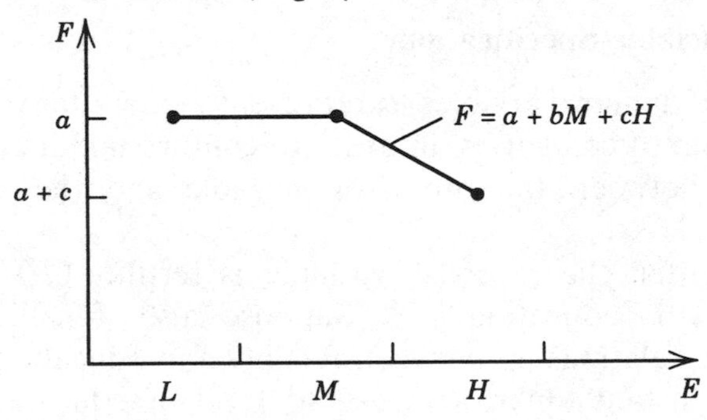

Figure 2.6. A nonlinear relationship between fertility (F) and education (E) for which a dummy variable specification would be appropriate. **Note:** In part b, L denotes low education, M denotes medium education, and H denotes high education. The fitted values of b and c are assumed to be $b = 0$ and $c < 0$. Values of F for low, medium, and high education are plotted at the midpoints of the three education intervals.

After fitting this model in the usual way, we can estimate fertility for each education category by substituting appropriate combinations of values of M and H in (2.29):

$$\text{Low education} \quad (M = 0, H = 0){:} \quad \hat{F} = a$$

$$\text{Medium education} \quad (M = 1, H = 0){:} \quad \hat{F} = a + b$$

$$\text{High education} \quad (M = 0, H = 1){:} \quad \hat{F} = a + c$$

Given our choice of cutting points for low, medium, and high education, we would find upon fitting (2.29) that $b = 0$ and $c < 0$, so that

$$\text{Low education:} \quad \hat{F} = a$$

$$\text{Medium education:} \quad \hat{F} = a$$

$$\text{High education:} \quad \hat{F} = a + c$$

These values, representing the mean values of F for low, medium, and high education, are plotted at the midpoints of the intervals for these three categories in Figure 2.6b. Although the two graphs in Figure 2.6 are by no means identical, Figure 2.6b nevertheless captures the piecewise-linear aspect of the original curve. If we wished, we could do better by further subdividing the education categories. For example, 16 dummy variables to represent $0, 1, 2, \ldots, 15$, and 16 + completed years of education would yield a piecewise-linear curve with 15 1-year straight-line segments that would capture quite well any single-year irregularities in the original curve.

2.8. GOODNESS OF FIT

We consider three measures of goodness of fit: the standard error of the estimate, s; the coefficient of determination, R^2 (and its square root, R, the multiple correlation coefficient); and the partial correlation coefficient.

2.8.1. Standard Error of the Estimate, *s*

Just as in bivariate linear regression, one measure of goodness of fit is σ, the standard deviation of the disturbances in the underlying population:

$$\sigma = \sqrt{\sigma^2} = \sqrt{\frac{\Sigma \varepsilon_i^2}{N}} \qquad \text{(1.24 repeated)}$$

When the calculation is done from sample data, it can be shown that

$$s^2 = \frac{\Sigma e_i^2}{n - k - 1} \qquad (2.30)$$

is an unbiased estimator of σ^2. In (2.30), k is the number of predictor variables and n is sample size. When $k = 1$, (2.30) reduces to (1.25). The square root of s^2, namely s, is called, as in bivariate linear regression, the *standard error of the estimate*.

2.8.2. Coefficient of Determination, R^2, and Multiple Correlation Coefficient, R

The estimated multiple regression model in (2.1b) can be written as

$$\hat{Y}_i = b_0 + b_1 X_{1i} + b_2 X_{2i} + \cdots + b_k X_{ki} \tag{2.31}$$

Starting from this model, we can decompose the total sum of squares into explained and unexplained components, just as in bivariate linear regression:

$$\boxed{\sum (Y_i - \bar{Y})^2 = \sum (Y_i - \hat{Y}_i)^2 + \sum (\hat{Y}_i - \bar{Y})^2} \tag{2.32}$$

This equation, the proof of which is omitted, is identical to (1.28) in Chapter 1. Equation (2.32) can be written more simply as

$$\boxed{\text{TSS} = \text{USS} + \text{ESS}} \tag{2.33}$$

The *coefficient of determination*, R^2, is defined as

$$\boxed{R^2 \equiv \frac{\text{ESS}}{\text{TSS}}} \tag{2.34}$$

The *multiple correlation coefficient*, R, is defined as the positive square root of R^2. (R is defined as always positive because there is no obvious way to assign a sign when some of the partial regression coefficients are positive and some are negative.) It can be shown that R is also the simple correlation between $\hat{Y}$ and Y:

$$\boxed{R = r_{\hat{Y}Y}} \tag{2.35}$$

When $k = 1$, so that there is only one predictor variable, R^2 reduces to r^2, and R reduces to r, except for sign because R is always positive.

Suppose that $k = 2$, so that the estimated regression model can be represented as a plane in 3-space. Then R^2 and R measure how closely the plotted points in 3-space fit the plane. If all the points fall precisely in the plane, then USS = 0, TSS = ESS, and $R^2 = R = 1$. If the points are distributed in a spherical cloud, then ESS = 0, TSS = USS, $b_1 = b_2 = 0$, and $R^2 = R = 0$. The extension to higher values of k is straightforward algebraically, but we can no longer portray it geometrically in a simple way.

2.8.3. Corrected R^2 and Corrected R

A problem with R^2 is that the addition of an irrelevant predictor variable to the multiple regression model will tend to increase R^2 even though the irrelevant predictor is in no way related to the other variables in the model. We can compensate for this increase by using *corrected* R^2 (also called *adjusted* R^2), denoted $\tilde{R}^2$:

$$\tilde{R}^2 \equiv \left(R^2 - \frac{k}{n-1}\right)\left(\frac{n-1}{n-k-1}\right) \tag{2.36}$$

A nice feature of $\tilde{R}^2$ is that it is related to the standard error of the estimate as

$$1 - \tilde{R}^2 = \frac{s^2}{s_Y^2} \tag{2.37}$$

where s^2 is defined by (2.30) and s_Y^2 by the square of s_Y in (1.35). Equation (2.37) says that $1 - \tilde{R}^2$ equals the ratio of the unexplained variance to the total variance in the response variable. (The operative word here is variance, not variation. Recall that $1 - R^2 = \text{USS}/\text{TSS}$.) (See Wonnacott and Wonnacott, 1979, p. 181; Theil, 1971, p. 178.)

Social scientists usually do not use $\tilde{R}^2$, because their surveys tend to be large (several thousands of cases). In the case of large sample sizes, $(k/n - 1)$ is close to zero and $(n-1)/(n-k-1)$ is close to unity in (2.36), so that to a close approximation, $\tilde{R}^2 = R^2$. Biomedical scientists, however, often deal with sample sizes of fewer than 50 cases, in which case $\tilde{R}^2$ is more commonly used.

2.8.4. Partial Correlation Coefficient

Suppose we wish to measure the correlation between Z and X while controlling for Y. This type of correlation coefficient is called a *partial correlation coefficient*, *denoted* $r_{ZX \cdot Y}$.

To calculate the partial correlation, we first regress each of the two variables of interest on the control variables. In our three-variable example, there is only one control variable, Y:

$$\hat{Z} = a + bY \tag{2.38}$$

$$\hat{X} = c + dY \tag{2.39}$$

We then remove the explanatory influence of Y on Z and X by forming the variables

$$Z' = Z - \hat{Z} \tag{2.40}$$

$$X' = X - \hat{X} \tag{2.41}$$

The residuals Z' and X' may be thought of as the unexplained parts of Z and X that are left when the effects of Y are removed. Then

$$\boxed{r_{ZX \cdot Y} = r_{Z'X'}} \tag{2.42}$$

where $r_{Z'X'}$ is the simple correlation between Z' and X'. $r_{ZX \cdot Y}$ may be thought of as a measure of how well the points (X', Z') fit the regression line $Z' = a + bX'$.

If we wish to control for another variable W as well as for Y, then (2.38) and (2.39) must each have an additional W term on the right, but otherwise the approach is the same. Equation (2.42) is unchanged, except that the partial correlation coefficient is now denoted $r_{ZX \cdot YW}$ on the left side of the equation.

The concept of partial r may be extended to partial R. Suppose, for example, that we have regressed Z on X and Y. We can compute an ordinary R for this regression. Now suppose that we wish to recompute R by controlling for a third variable W. We proceed as before. First we regress Z, X, and Y on W:

$$\hat{Z} = a + bW \tag{2.43}$$

$$\hat{X} = c + dW \tag{2.44}$$

$$\hat{Y} = e + fW \tag{2.45}$$

We then remove the explanatory influence of W by forming the variables

$$Z' = Z - \hat{Z} \tag{2.46}$$

$$X' = X - \hat{X} \tag{2.47}$$

$$Y' = Y - \hat{Y} \tag{2.48}$$

Then partial R, controlling for W, is calculated as ordinary R for the regression

$$\hat{Z}' = g + hX' + iY' \tag{2.49}$$

We denote this partial R as $R_{ZXY \cdot W}$. Computer programs for partial r are contained in such standard statistical packages as SPSS and BMDP. The user specifies which variable is the response variable, which variable is the predictor variable, and which variables are the control variables. The computer does the rest. However, neither SPSS nor BMDP provides programs for partial R, except for SPSS, which calculates partial R in the subprogram for multiple classification analysis, to be discussed in the next chapter. The lack of a special program is not a problem, because partial R can be programmed in a few simple steps in any statistical package with multiple regression. (See Appendix A.) In the above example, one would simply create new variables Z', X', and Y' as in (2.46)–(2.48), and then obtain R by fitting the model in (2.49).

2.9. STATISTICAL INFERENCE

2.9.1. Hypothesis Testing, Confidence Intervals, and *p* Values for a Single Regression Coefficient

How can we generalize from b_k to β_k? To proceed, we must first determine the sampling distribution of b_k. As in bivariate linear regression, this sampling distribution is derived theoretically. In this book, we simply accept the derivation on faith without going into mathematical details. The estimated standard deviation of the sampling distribution is denoted s_{b_k} and is called the *standard error of* b_k. Once we know s_{b_k}, we undertake hypothesis testing and construct confidence intervals using the same procedures and formulac givcn for bivariate linear regression in Chapter 1. The only differences are that

b_k is substituted for b, and the degrees of freedom associated with t are d.f. $= n - k - 1$ instead of $n - 2$, where k is the number of predictor variables. Thus statistical inference from b_k to β_k in multiple regression is done in the same way as statistical inference from b to β in bivariate linear regression. In this regard, as in most other regards, bivariate linear regression may be viewed as a special case of multiple regression. In practice, computer programs for multiple regression routinely print out standard errors, t values (which again may be treated as Z values if d.f. is larger than about 120), and p values for the intercept and each coefficient, so that the computer does most of the work. The printed t values and p values assume that the null hypothesis is $\beta_j = 0$ for $0 \leq j \leq k$.

2.9.2. Testing the Difference Between Two Regression Coefficients, β_i and β_j

Suppose that we are looking at the effect of occupation (white collar, blue collar, farm) on fertility (number of children ever born), and suppose that the estimated model is

$$\hat{F} = a + bW + cB \qquad \text{(2.11 repeated)}$$

where F is fertility, W is 1 if white collar and 0 otherwise, B is 1 if blue collar and 0 otherwise, and farm is the reference category. The usual statistical tests on the coefficient b will tell us whether the fertility effect of being white collar, relative to farm, differs significantly from zero. Likewise the usual tests on c will tell us whether the fertility effect of being blue collar, relative to farm, differs significantly from zero. But how can we test (without redefining the reference category and rerunning the regression) whether the effect of being white collar, relative to blue collar, differs significantly from zero?

To answer this question, we note first that the effect on F of being white collar, relative to blue collar, is $b - c$. (See earlier discussion of equations (2.11) and (2.13).) Therefore, we must test whether $b - c$ differs significantly from zero. The first step is to compute the variance of $b - c$ from the formula[3]

$$\text{Var}(b - c) = \text{Var}(b) + \text{Var}(c) - 2\,\text{Cov}(b, c) \qquad (2.50)$$

[3]Equation (2.50) is a specific instance of the more general formula

$$\text{Var}\left(\sum_i a_i X_i\right) = \sum a_i^2\,\text{Var}(X_i) + 2\sum_{j>i} a_i a_j\,\text{Cov}(X_i, X_j)$$

Standard computer programs for multiple regression print out the standard errors of b and c, which, when squared, yield Var(b) and Var(c), and they also contain an option for printing out Cov(b, c). This covariance appears as an entry in a *regression coefficient covariance matrix*. (In SAS, for example, the table labeled "Parameter Estimates" can be supplemented by an additional table labeled "Covariance of Estimates.") This is a table that has the variables listed both down the side and across the top as row and column labels, and in which off-diagonal cell entries are covariances for pairs of regression coefficients. The diagonal entries are variances of regression coefficients, which, as already mentioned, can be obtained alternatively by squaring the standard errors of the coefficients.

Cov(b, c), like Var(b) and Var(c), is to be thought of in the context of repeated samples from the underlying population. For each such sample, a regression is fitted and a pair of values (b, c) is obtained. The covariance of b and c is then calculated from the set of pairs (b, c) over repeated samples. In practice, of course, we have only one sample; Cov(b, c), like Var(b) and Var(c), is derived from statistical theory, not by actually taking repeated samples.

The square root of Var($b - c$) is the standard error of $b - c$, which may be used to test whether the value of $b - c$ for our particular sample differs significantly from zero. Such a test is equivalent to a test of whether b differs significantly from c. The test of whether $b - c$ differs significantly from zero is a standard t test with $n - k - 1$ degrees of freedom. The t value is calculated as

$$t = \frac{b - c}{\sqrt{\mathrm{Var}(b - c)}} \tag{2.51}$$

2.9.3. Testing Effects When There Is Interaction

Suppose that our estimated model is

$$\hat{F} = a + bI + cE + dIE \qquad \text{(2.14 repeated)}$$

where F is fertility, I is income, and E is number of completed years of education. The effect of a one-unit increase in income on fertility is $b + dE$. [See earlier discussion of equations (2.14)–(2.16).]

Note that when there is interaction, coefficients are not the same as effects. The coefficient of I is b, but the effect of I is $b + dE$. We already know how to test whether the coefficient is statistically significant. How do we test whether the effect is significant?

For simplicity, let us denote $b + dE$ as q. We must test whether q differs significantly from zero. To do this, we form the t statistic, $t = q/s_q$, and then proceed as usual to test whether q differs significantly from zero.

The standard error s_q is calculated as $s_q = \sqrt{\text{Var}(b + dE)}$. From the general formula in footnote 3 above, we obtain $\text{Var}(b + dE)$ as

$$\text{Var}(b + dE) = \text{Var}(b) + E^2 \text{Var}(d) + 2E \text{Cov}(b, d) \quad (2.52)$$

In this formulae, E is treated as a constant, because it is preset to a particular value. The calculation formula for t is then

$$t = \frac{b + dE}{\sqrt{\text{Var}(b) + E^2 \text{Var}(d) + 2E \text{Cov}(b, d)}} \quad (2.53)$$

We may now perform a standard t test with d.f. $= n - k - 1$.

In (2.53), the effect $b + dE$ can be significant even when the coefficients b and d are not significant. Therefore, when interaction is present, it is not enough to test for significance of coefficients. One must also test for significance of effects. The effect, $b + dE$, may be significant at some values of E and nonsignificant at other values of E.[4]

2.9.4. Testing Effects When There Is a Nonlinearity

Suppose that our estimated model is

$$\hat{F} = a + bE + cE^2 \quad \text{(2.26 repeated)}$$

[4] By way of comparison, suppose a simple model without interaction, $\hat{F} = a + bE$, where b, representing the effect on F of a one-unit increase in E, is nonsignificant. Might the effect of a three-unit increase in E be significant, even though the effect of a one-unit increase in E is not? For the three-unit increase,

$$t = \frac{3b}{\sqrt{\text{Var}(3b)}} = \frac{3b}{3\sqrt{\text{Var}(b)}} = \frac{b}{\sqrt{\text{Var}(b)}}$$

Because of cancellation, the t value for $3b$ is the same as the t value for b. Therefore, if the effect of a one-unit increase in E is nonsignificant, the effect of a three-unit increase in E is also nonsignificant.

In contrast, when there is interaction, as in (2.53), the t value depends on E because there is no cancellation of terms in E between the numerator and denominator of (2.53).

where F is fertility and E is number of completed years of education. The (instantaneous) effect of E on F is $b + 2cE$. [See earlier discussion following equation (2.28).]

Here again, the coefficient of E is not the same as the effect of E. For simplicity, let us denote $b + 2cE$ as q. Our strategy is to calculate $t = q/s_q$ and then perform a standard t test of whether q differs significantly from zero.

From the general formula in footnote 3, we obtain $\text{Var}(b + 2cE)$ as

$$\text{Var}(b + 2cE) = \text{Var}(b) + 4E^2\,\text{Var}(c) + 4E\,\text{Cov}(b, c) \quad (2.54)$$

The calculation formula for t is then

$$t = \frac{b + 2cE}{\sqrt{\text{Var}(b) + 4E^2\,\text{Var}(c) + 4E\,\text{Cov}(b, c)}} \quad (2.55)$$

We can now perform a standard t test with d.f. $= n - k - 1$.

The effect $b + 2cE$ can be significant even when the coefficients b and c are not significant. Therefore, when nonlinearities are present, one must test for significance of effects as well as significance of coefficients. The effect $b + 2cE$ may be significant at some values of E and nonsignificant at other values of E.

2.9.5. The ANOVA Table

The so-called "ANOVA table" (ANOVA is an acronym for analysis of variance) is shown in Table 2.1. In the table, k degrees of freedom are associated with ESS, and the *explained variance*, V_E, is defined as ESS$/k$. Similarly, $n - k - 1$ degrees of freedom are associated with

TABLE 2.1. Analysis of Variance (ANOVA) Table for Multiple Regression

Source of Variation	Variation	d.f.	Variance	F Ratio
Explained	ESS	k	$\dfrac{\text{ESS}}{k} \equiv V_E$	
Unexplained	USS	$n - k - 1$	$\dfrac{\text{USS}}{n - k - 1} \equiv V_U$	
Total	TSS	$n - 1$		
				$\dfrac{V_E}{V_U}$

USS, and the *unexplained variance*, V_U, is defined as USS$/(n - k - 1)$. The *F ratio* is defined as

$$F \equiv \frac{V_E}{V_U} \tag{2.56}$$

The better the fit of the multiple regression model to the sample data, the larger the value of V_E*, the smaller the value of* V_U*, and the larger the value of* F. Note that F is always positive.

To the nonstatistician, the ANOVA table may seem rather strange. In particular, F may seem rather unfamiliar and mysterious. *The point to be appreciated about F is that it has a known sampling distribution, which is given in Table B.3 (Appendix B) in the form of selected critical points. Because F has a known sampling distribution, we can use it for statistical inference from sample to underlying population*, as illustrated in the next section.

The ANOVA table and the F ratio illustrate the theoretical statistician's general strategy for finding a way to generalize from sample to population: If, for a sample statistic of interest, one cannot determine an appropriate sampling distribution directly, first transform the statistic into some other statistic for which a sampling distribution can be derived, and then use this new statistic for statistical inference. The omnibus F test in the next section illustrates this strategy.

2.9.6. The Omnibus F Test of the Hypothesis $\beta_1 = \beta_2 = \cdots = \beta_k = 0$

The *omnibus F test* is a *simultaneous hypothesis test* of whether all the coefficients β_j, $j = 1, 2, \ldots, k$, are simultaneously zero.

Suppose that H_0: $\beta_1 = \beta_2 = \cdots = \beta_k = 0$ is true. Then, in the underlying population, ESS and V_E are both zero, so that $F = 0$. In the sample, however, chances are that $F > 0$. Under the assumption that H_0 is true, the sampling distribution of F, in the form of F critical points, is given in Table B.3 (Appendix B). In this table, d.f. $= k$ for the numerator of F, and d.f. $= n - k - 1$ for the denominator of F. If the sample value of F exceeds the critical point at, say, the 5 percent level of significance, we reject H_0. When we do so, we run a 5 percent risk of falsely rejecting a true hypothesis.

To illustrate, let us suppose that $k = 5$, $n - k - 1 = 40$, and $F = 4.32$ for our particular sample. Thus d.f. $= 5$ for the numerator of

F, and d.f. = 40 for the denominator of F. We consult the table of F critical points in Table B.3 (Appendix B). For d.f. = (5, 40), we see from the table that the critical point for a test at the 5 percent level of significance is $F = 2.45$. Our value of $F = 4.32$ is larger than this, so we reject H_0: $\beta_1 = \beta_2 = \cdots = \beta_k = 0$ at the 5 percent level of significance. Because the critical point for a 1 percent level of significance is 3.51, we can also reject H_0 at the 1 percent level. However, the critical point for the 0.1 percent level is $F = 5.13$, which is larger than our observed F of 4.32, so we cannot reject H_0 at the 0.1 percent level of significance. Another way to report the results of this test is to write $.001 < p < .01$.

Were our table of F critical points complete, we could calculate an exact p value, or observed level of significance. But such a table would be so voluminous as to be impractical. Fortunately, the ANOVA table and the exact (to within rounding error) p value for F can be printed out by most multiple regression computer programs, in which case we need not bother consulting a table of F critical points. The p value for F is printed out along with the ANOVA table.

2.9.7. Test of the Hypothesis That Some of the β_j Are Zero

Suppose that our null hypothesis is that $\beta_j = 0$ for a subset of m predictors. Let us order the k predictors so that the subset of m predictors comes last.

First we run the regression with the first $k - m$ predictors, omitting the last m predictors, and calculate R^2. Then we rerun the regression with all k predictors and recalculate R^2. Denoting the increase in R^2 as ΔR^2, we then form the F ratio

$$F = \frac{\dfrac{\Delta R^2}{m}}{\dfrac{1 - R^2}{n - k - 1}} \tag{2.57}$$

where R^2 is calculated from the regression with all k predictors. If H_0: $\beta_j = 0$, $j = k - m + 1, \ldots, k$, is true, this F ratio has an F distribution with d.f. = $(m, n - k - 1)$. We then compare the observed value of F with the appropriate F critical point at, say, the 5 percent level of significance in Table B.3 (Appendix B). If our observed F exceeds the cutoff value, we reject H_0: $\beta_{k-m+1} = \beta_{k-m+2} = \cdots = \beta_k = 0$.

In setting up this F test, the order of the variables must be determined beforehand by theory, not by the data themselves as in stepwise regression (see following section). The F test is invalid if stepwise regression is used to determine the order of importance of the predictor variables.

2.10. STEPWISE REGRESSION

Sometimes, when there is a large number of predictor variables, the computer is programmed to introduce predictor variables one at a time in order to determine the sequence that makes R^2 climb the fastest. Only the predictors that are the most powerful by this criterion are retained in the final model.

This is done in the following way: First, the program runs a series of bivariate regressions, one for each predictor variable. It retains the predictor variable from the regression with the highest R^2. Then the program runs a second series of regressions with two predictor variables, each of which includes the predictor variable retained from the first series, plus one of the remaining predictor variables. From these remaining predictor variables, the program retains the one from the multiple regression with the highest R^2 in the second series. Then the program runs a third series of regressions with three predictor variables, each of which includes the two predictor variables retained from the first two series of regressions, plus one of the remaining predictor variables. From these remaining predictor variables, the program retains the one from the multiple regression with the highest R^2 in the third series. And so on until the desired number of retained predictor variables is reached, or until there is no additional variable that improves R^2 significantly by the F test described in the previous section. The final regression is the one that includes these retained predictor variables and no others.

If one wishes to drop only one or two predictors, one may alternatively use the *method of backward deletion*, which deletes predictors one at a time. For example, to delete the weakest predictor, the program first runs a multiple regression that includes all the predictor variables and computes an R^2 for this regression. Then the program runs a series of multiple regressions, each of which deletes one of the predictor variables. An R^2 is computed for each of these regressions. The weakest predictor is chosen as the deleted predictor variable corresponding to the regression with the highest R^2 in this series. This latter regression is retained as the final regression.

Although stepwise regression (forward or backward) may be useful for exploring the data, it is a brute-force technique. In letting the computer rather than theory dictate the form of the model, one can end up with theoretically important variables dropped from the model. In general, it seems preferable that model specification be guided by theory instead of some mechanical rule.

Stepwise regression is additionally frowned on by many statisticians because it violates the logic of hypothesis testing. Instead of testing a theory, one typically concocts an *ad hoc* explanation that seems consistent with a model arrived at mechanically by stepwise regression. Statistical tests are then inappropriate and invalid.

2.11. ILLUSTRATIVE EXAMPLES

2.11.1. Example 1

We consider next an example that illustrates numerically many of the points made in this chapter. The example is taken from the 1974 Fiji Fertility Survey, which was part of the World Fertility Survey. Individual record data from this survey are in the public domain, are available on computer tape, and therefore provide a convenient source of constructed examples used throughout the rest of this book.

Suppose that we wish to assess the effects of age, education (in completed years), residence (urban or rural), and ethnicity (Indian or indigenous Fijian) on fertility (number of children ever born), for ever-married women aged 20–49 at the time of the survey.[5] The variables are

F: fertility (number of children ever born)
A: woman's age
E: education (number of completed years)
U: residence (1 if urban, 0 otherwise)
I: ethnicity (1 if Indian, 0 otherwise)

Because we expect that the effect of age on number of children ever born tends to decrease as age increases (for various reasons the pace

[5]Indians, who originally came to Fiji as plantation labor, constituted somewhat over half of Fiji's population in 1974. Fijians constituted somewhat less than half. We removed from the sample the scattering of Caucasians and other racial or ethnic groups in order to simplify our constructed examples, both in this chapter and in subsequent chapters.

of childbearing slows down with age), it makes sense to include an age-squared term in the model. If we also assume that the effect of residence on fertility varies by ethnicity, then the model should also include an interaction term. (As it turns out, this term is not statistically significant, but we retain it anyway for purposes of illustration.) The form of the estimated model is

$$\hat{F} = a + bA + cA^2 + dE + eU + fI + gUI \tag{2.58}$$

When fitted to Fiji data, (2.58) becomes

$$\hat{F} = -6.499 + .477A - .004A^2 - .099E - .491U + .312I + .006UI \tag{2.59}$$

All coefficients differ significantly from zero at the 5 percent level (two-sided $p < .05$), except for the coefficient of the UI term. Coefficients and effects, which are derived by formulae derived in earlier parts of this chapter, are presented in Table 2.2.

The regression coefficient covariance matrix is given in Table 2.3. Using information from this table, we can calculate the t value for the

TABLE 2.2. Coefficients and Effects for the Estimated Model $\hat{F} = a + bA + cA^2 + dE + eU + fI + gUI$: Ever-Married Women Aged 20–49 in the 1974 Fiji Fertility Survey[a]

Predictor Variable	Coefficient	Effect
Intercept	−6.499* (.610)	—
Age		$.477 + 2(-.004)A = .477 - .008A$
A	.477* (.037)	—
A^2	−.004* (.001)	—
Education		
E	−.099* (.012)	−.099
Residence × ethnicity		
U	−.491* (.111)	$-.491 + .006I$
I	.312* (.092)	$.312 + .006U$
UI	.006 (.144)	—

[a]Numbers in parentheses are standard errors of coefficients. An asterisk after a coefficient denotes a two-sided $p < .05$ for H_0: $\beta_i = 0$. The entry for the effect of A uses the formula for the instantaneous effect of A. Significance is not indicated in the effect column, because in all but one case the t value depends on either A, U, or I, particular values of which are not specified.

TABLE 2.3. Regression Coefficient Covariance Matrix for the Estimated Model $\hat{F} = a + bA + cA^2 + dE + eU + fI + gUI$: Ever-Married Women Aged 20–49 in the 1974 Fiji Fertility Survey

	b	*c*	*d*	*e*	*f*	*g*
b	.0013					
c	−.0000	.0000				
d	.0000	.0000	.0001			
e	.0000	.0000	−.0001	.0123		
f	.0001	.0000	.0005	.0023	.0085	
g	.0000	.0000	−.0002	−.0121	−.0075	.0206

effect of age, which is $b + 2cA$, as

$$\begin{aligned} t &= \frac{b + 2cA}{\sqrt{\mathrm{Var}(b) + 4A^2\,\mathrm{Var}(c) + 4A\,\mathrm{Cov}(b,c)}} \\ &= \frac{.477 - .008A}{\sqrt{(.037)^2 + 4A^2(.001)^2 + 4A(-.0000)}} \end{aligned} \tag{2.60}$$

In (2.60), we have obtained variances by squaring standard errors of coefficients instead of obtaining them from the diagonal entries of the regression coefficient covariance matrix. The covariance term, which is zero to four decimal places, is very small in this case and can be ignored. If $A = \bar{A} = 32.36$, then $t = 39.25$. Because 39.25 is much larger than 1.96, the effect of age is significant at the 5 percent level at $A = 32.36$.

Similarly, the t value for the effect of U is

$$\begin{aligned} t &= \frac{e + gI}{\sqrt{\mathrm{Var}(e) + I^2\,\mathrm{Var}(g) + 2I\,\mathrm{Cov}(e,g)}} \\ &= \frac{-.491 + .006I}{\sqrt{(.111)^2 + I^2(.144)^2 + 2I(-.0121)}} \end{aligned} \tag{2.61}$$

If $I = 0$, then $e + gI = -.491$ and $t = -4.42$. That is, for Fijians, the effect of being urban, relative to rural, is to lower fertility by .491 child, and the effect is significant at the 5 percent level, because $|t| > 1.96$. If $I = 1$, then $e + gI = -.485$ and $t = -5.15$. That is, for Indians, the effect of being urban, relative to rural, is to lower fertility by .485 child, and the effect is significant.

TABLE 2.4. ANOVA Table for the Estimated Model $\hat{F} = a + sbA + cA^2 + dE + e^i + fI + gUI$: Ever-Married Women Aged 20–49 in the 1974 Fiji Fertility Survey[a]

Source of Variation	Variation	d.f.	Variance	F Ratio
Explained	15,477	6	2579.50	
Unexplained	22,126	4488	4.93	
Total	37,603	4494		
				523

[a]The observed level of significance for this model, based on $F = 523$ with d.f. = (6, 4488), is $p = .0001$. Because $p < .05$, we reject the hypothesis that the underlying population coefficients corresponding to b, c, d, e, f, and g are all simultaneously zero. Also printed out with the ANOVA table are values of R^2 and $\tilde{R}^2$, which are .4116 and .4108 respectively.

The ANOVA table for this model is shown in Table 2.4. The value of F is highly significant, so we reject the hypothesis that all the coefficients are simultaneously zero. Values of R^2 and $\tilde{R}^2$ are .4116 and .4108, respectively; close agreement is expected, because the sample size is large, 4,495 cases.

2.11.2. Example 2

Let us consider the estimated model

$$\hat{F} = a + bA + cM + dH + eU + fI + gUI \qquad (2.62)$$

for women aged 35–49 in the Fiji Fertility Survey. This is the same as the model in the previous example, except that age is linear (the quadratic term A^2 is unnecessary because of the more truncated age range) and education is represented as low, medium, or high education, with low education as the reference category.

Alternatively, we can rewrite (2.62) with medium instead of low education as the reference category for education:

$$\hat{F} = a' + bA + iL + jH + eU + fI + gUI \qquad (2.63)$$

By the same logic employed earlier in Section 2.4.2, $a' = a + c$, $i = -c$, and $j = d - c$.

If we fit (2.62) and (2.63) separately to the Fiji data, we get the results shown in Table 2.5. The model with low education as the reference category shows that the coefficient of H is significant but the coefficient of M is not. The model with medium education as the

TABLE 2.5. Regressions of Fertility on Age (A), Education (L, M, H), Residence (U), and Ethnicity (I): Ever-Married Women Aged 35–49 in the 1974 Fiji Fertility Survey[a]

	Model with Low Education as Reference Category		Model with Medium Education as Reference Category	
Variable	Coefficient	S.E.	Coefficient	S.E.
Intercept	−.040	.771	−.311	.743
A	.148*	.018	.148*	.018
L	—	—	.271	.186
M	−.271	.186	—	—
H	−.613*	.220	−.343	.219
U	−1.129*	.239	−1.129*	.239
I	.687*	.201	.687*	.201
UI	.296	.310	.296	.310

[a]An asterisk after a coefficient indicates that the coefficient differs significantly from zero at the 5 percent level.

reference category shows that neither the coefficient of L nor the coefficient of H is significant. Judging solely from the second model, we might come to the erroneous conclusion that education has no significant effects on fertility. *It is necessary to check for significant fertility differences between each possible pair of education categories.* This can be done either by rerunning the regression or by deriving one regression from the other by the approach discussed earlier in Section 2.4.2.

Suppose that we have run (2.62) and wish to know whether the effect on F of high education, relative to medium education, is significant, without rerunning the regression. As already mentioned, we calculate the effect on F of H, relative to M, as $d - c$, which, from the first column of Table 2.5, is $-.613 - (-.271) = -.342$. This value agrees with the value of $-.343$ in the third column of Table 2.5 to within rounding error. We can calculate a t value for $d - c = -.342$ as

$$t = \frac{d-c}{\sqrt{\mathrm{Var}(d-c)}} = \frac{d-c}{\sqrt{\mathrm{Var}(d) + \mathrm{Var}(c) - 2\,\mathrm{Cov}(c,d)}}$$

$$= \frac{-.342}{\sqrt{(.220)^2 + (.186)^2 - 2(.018)}} = -1.58 \qquad (2.64)$$

(The regression coefficient covariance matrix from which .018 is obtained is not shown.) This value agrees to within rounding error with $t = -.343/.219 = -1.57$, calculated from the coefficient and standard error of H in the last two columns of Table 2.5. Because the absolute value of t, 1.58, is less than 1.96, the effect on F of high education, relative to medium education, is not significant at the 5 percent level.

2.12. FURTHER READING

The same recommendations for further reading on bivariate linear regression, made at the end of Chapter 1, apply to multiple regression as well.

3

MULTIPLE CLASSIFICATION ANALYSIS

Multiple classification analysis (MCA) is most easily explained as multiple regression with dummy variables. *In basic MCA, the response variable is a quantitative variable, and the predictor variables are categorical variables, represented by dummy variables.*

Recall from introductory statistics courses that also in simple analysis of variance, the response variable is quantitative, and the predictor variables are categorical. It turns out that MCA with one categorical predictor variable is equivalent to one-way analysis of variance; MCA with two categorical predictor variables is equivalent to two-way analysis of variance; and so on.

Control variables may be added to the MCA model. When quantitative variables, in addition to categorical variables, are included among the predictor variables, MCA is equivalent to what is commonly called *analysis of covariance*.

MCA can be extended beyond multiple regression. In later chapters we shall extend it to logistic regression, multinomial logistic regression, and hazard models.

3.1. THE BASIC MCA TABLE

The discussion of MCA in this chapter continues the use of examples from the 1974 Fiji Fertility Survey. Suppose, for the Fiji data, that we wish to assess the effect of education (low, medium, and high), residence (urban, rural), and ethnicity (Indian, indigenous Fijian) on fertility (number of children ever born). We have a quantitative response variable and three categorical predictor variables, denoted as follows:

F: fertility (number of children ever born)
M: 1 if medium education, 0 otherwise
H: 1 if high education, 0 otherwise
U: 1 if urban, 0 otherwise (i.e., rural)
I: 1 if Indian, 0 otherwise (i.e., Fijian)

Low education is defined as 0–4 completed years of education, medium education as 5–7 years, and high education as 8 or more years. Throughout this chapter, we shall restrict consideration to ever-married women aged 40–49 at the time of the survey, in order to obviate the need for a control for woman's age.

The set-up for the basic MCA table is shown in Table 3.1. *This table contains two sets of results, unadjusted and adjusted.* (Sometimes "unadjusted" and "adjusted" are called "gross" and "net".) "Unadjusted" means "without controls," and "adjusted" means "with controls." In Table 3.1, the unadjusted and adjusted results are in the form of predicted fertility values, obtained from regressions.

3.1.1. Unadjusted Values

Formulae for calculating unadjusted values of fertility for categories of the predictor variable are given in the unadjusted fertility column of Table 3.1. The three cell entries for education are derived from the regression of fertility on education:

$$\hat{F} = a' + b'M + c'H \tag{3.1}$$

The two cell entries for regression are derived from the regression of

TABLE 3.1. Set-up for MCA Table of the Effects of Education, Residence, and Ethnicity on Fertility[a]

Predictor Variable	n	Unadjusted Fertility	Unadjusted R	Adjusted Fertility	Adjusted R
Education			R_{FMH}		$R_{FMH \cdot UI}$
Low	n_L	a'		$a + d\overline{U} + e\overline{I}$	
Medium	n_M	$a' + b'$		$a + b + d\overline{U} + e\overline{I}$	
High	n_H	$a' + c'$		$a + c + d\overline{U} + e\overline{I}$	
Residence			R_{FU}		$R_{FU \cdot MHI}$
Urban	n_U	$a'' + d''$		$a + d + b\overline{M} + c\overline{H} + e\overline{I}$	
Rural	n_R	a''		$a + b\overline{M} + c\overline{H} + e\overline{I}$	
Ethnicity			R_{FI}		$R_{FI \cdot MHU}$
Indian	n_I	$a''' + e'''$		$a + e + b\overline{M} + c\overline{H} + d\overline{U}$	
Fijian	n_F	a'''		$a + b\overline{M} + c\overline{H} + d\overline{U}$	
All predictor variables (full model)					R

[a]The underlying regressions are:

Unadjusted: Education: $\hat{F} = a' + b'M + c'H$

Residence: $\hat{F} = a'' + d''U$

Ethnicity: $\hat{F} = a''' + e'''I$

Adjusted: $\hat{F} = a + bM + cH + dU + eI$

The tables in this chapter all refer to ever-married women aged 40–49 in the 1974 Fiji Fertility Survey, which was part of the World Fertility Survey.

fertility on residence:

$$\hat{F} = a'' + d''U \tag{3.2}$$

The two cell entries for ethnicity are derived from the regression of fertility on ethnicity:

$$\hat{F} = a''' + e'''I \tag{3.3}$$

The reasons for the primes, double primes, triple primes, and letters chosen to denote coefficients have to do with parallel notation for the underlying regression for the adjusted effects, as will be explained shortly.

From these three equations, formulae for computing unadjusted values of fertility for categories of the predictor variables are derived by setting the predictor variables to appropriate combinations of ones and zeros. For example, the formulae for unadjusted fertility values for low, medium, and high education in Table 3.1 are obtained by setting (M, H) alternatively to $(0,0)$, $(1,0)$, and $(0,1)$ in equation (3.1). The formulae for categories of residence are obtained by setting U alternatively to 1 and 0 in (3.2). The formulae for categories of ethnicity are obtained by setting I alternatively to 1 and 0 in (3.3).

Alternatively but equivalently, one may derive unadjusted values of fertility for categories of the predictor variables by computing mean fertility for each category. For example, if one computes mean fertility for those with medium education, the result is identical to that obtained from the formula $a' + b'$. If one computes mean fertility for urban, the result is identical to that computed from the formula $a'' + d''$.[1]

Because this alternative approach to calculating unadjusted fertility values is simpler, why bother with the regressions in (3.1)–(3.3)? The reasons are (1) we need the regressions to interpret the measure of goodness of fit, R, to be discussed shortly, and (2) the regressions are useful for understanding the logic of the comparison between unadjusted and adjusted values.

The unadjusted fertility values in Table 3.1 are not the same as the unadjusted fertility *effects* of the predictor variables. *The unadjusted fertility effects are computed as differences between the unadjusted fertility values for pairs of categories of each predictor variable.* For example, the effect of medium education on fertility, relative to low education, is computed as $(a' + b') - a' = b'$, which is the coefficient of M in (3.1). The effect on fertility of being urban, relative to rural, is computed as $(a'' + d'') - a'' = d''$, which is the coefficient of U in (3.2). These results are consistent with the meaning of "effect" in the earlier discussion of multiple regression with dummy variables in Chapter 2.

[1]It is easy to see why this equivalence exists. In (3.2), for example, the regression line must pass through the mean of the F values at $U = 0$, and through the mean of the F values at $U = 1$. But the mean of the F values at $U = 0$ is simply mean fertility for rural, and the mean of the F values at $U = 1$ is simply mean fertility for urban. Similarly, in (3.1) the regression plane must pass through the means of the F values at $M = 0$ and $M = 1$ when H is held constant, and through the means of the F values at $H = 0$ and $H = 1$ when M is held constant.

3.1.2. Adjusted Values

In Table 3.1, formulae for calculating adjusted values of fertility for categories of the predictor variables are given in the adjusted fertility column. These formulae are derived from the underlying regression of fertility on all the predictor variables:

$$\hat{F} = a + bM + cH + dU + eI \tag{3.4}$$

As mentioned earlier, "adjusted" means "with statistical controls." In the first panel of Table 3.1, which tabulates fertility by education, residence and ethnicity are the control variables. In the second panel, which tabulates fertility by residence, education and ethnicity are the control variables. In the third panel, which tabulates fertility by ethnicity, education and residence are the control variables. *More generally, when we consider adjusted fertility tabulated by one predictor variable, all the other predictor variables are the control variables. Thus the set of control variables changes as we move down the table.*

The formulae for computing adjusted values of fertility for categories of a particular predictor variable are derived from (3.4) by setting the predictor variables to appropriate combinations of ones, zeros, and mean values. *Statistical controls are introduced by holding the control variables constant at their mean values in the entire sample.*

The formulae for adjusted values of fertility for categories of education in Table 3.1 are obtained by setting (M, H) alternatively to $(0, 0)$, $(1, 0)$, and $(0, 1)$ in the equation

$$\hat{F} = a + bM + cH + d\bar{U} + e\bar{I} \tag{3.5}$$

The formulae for categories of residence are obtained by setting U alternatively to 1 and 0 in the equation

$$\hat{F} = a + dU + b\bar{M} + c\bar{H} + e\bar{I} \tag{3.6}$$

The formulae for categories of ethnicity are obtained by setting I alternatively to 1 and 0 in the equation

$$\hat{F} = a + eI + b\bar{M} + c\bar{H} + d\bar{U} \tag{3.7}$$

Equations (3.5)–(3.7) are derived from equation (3.4) by setting whichever variables are considered the control variables to their mean values in the entire sample (which in this context is all ever-married women aged 40–49 at the time of the survey).

Note that the adjusted fertility value for the reference category of education (i.e., low education) is $a + d\bar{U} + e\bar{I}$, not a. Similarly, the adjusted fertility value for the reference category of residence (i.e., rural) is $a + b\bar{M} + c\bar{H} + e\bar{I}$. The adjusted fertility value for the reference category of ethnicity (i.e., Fijian) is $a + b\bar{M} + c\bar{H} + d\bar{U}$. The intercept a by itself is simply the predicted value of F for persons for whom all the predictor variables equal zero.

Adjusted fertility effects are computed as differences between adjusted fertility values. For example, the adjusted fertility effect of medium education, relative to low education, is computed as $(a + b + d\bar{U} + e\bar{I}) - (a + d\bar{U} + e\bar{I}) = b$. The adjusted fertility effect of being urban, relative to rural, is computed as $(a + d + b\bar{H} + c\bar{M} + e\bar{I}) - (a + b\bar{H} + c\bar{M} + e\bar{I}) = d$. The adjusted effect of being Indian, relative to Fijian, is computed as $(a + e + b\bar{H} + c\bar{M} + d\bar{U}) - (a + \bar{H} + c\bar{M} + d\bar{U}) = e$. Again these results are consistent with the meaning of "effect" in the earlier discussion of multiple regression with dummy variables in Chapter 2.

Although controls are normally introduced by holding the control variables constant at their mean values in the entire sample, other values can also be used. For example, consider again the fertility effect of medium education, relative to low education. By holding U and I constant at $\bar{U}$ and $\bar{I}$, the terms in $\bar{U}$ and $\bar{I}$ cancel out when computing the difference $(a + b + d\bar{U} + e\bar{I}) - (a + d\bar{U} + e\bar{I}) = b$. *Because of the cancellation, it doesn't matter how we interpret $\bar{U}$ and $\bar{I}$; any values will do in the sense that the difference will still be b.* For example, under some circumstances we might want to reinterpret $\bar{U}$ and $\bar{I}$ as the mean values of U and I for Fijians instead of the entire sample ($\bar{I} = 0$ for Fijians.) *The reinterpretation of $\bar{U}$ and $\bar{I}$ affects the levels of the adjusted fertility values but not the adjusted fertility effects, as measured by pairwise differences between the adjusted fertility values.*

In MCA we are interested mainly in assessing effects, which are measured by differences between fertility values and therefore unaffected by how we set the levels of the control variables. We set the control variables to whatever levels are convenient for analysis. In the absence of a reason to do otherwise, we normally choose these values to be the means of the control variables in the entire sample.

In some cases the use of mean values may seem somewhat artificial. For example, it may seem unrealistic to use $\bar{U}$ as the level of the control variable U, because no individual can assume this particular value. (Individuals are either urban or rural—that is, $U = 1$ or $U = 0$

—but not in between.) Nevertheless, we usually view $\overline{U}$ (the proportion urban) as convenient for the purpose of introducing statistical controls.

Computer programs for MCA automatically set the control variables at their mean values in the entire sample. If one wishes to use values other than mean values, it is necessary to run the underlying multiple regressions and then construct the unadjusted and adjusted fertility values from these regressions using the approach illustrated in Table 3.1. Those who have some experience with computers can shorten the work (and lessen the risk of making computational errors on a calculator) by downloading the regression results into a spreadsheet program such as LOTUS and doing the remainder of the calculations in LOTUS.

3.1.3. Unadjusted and Adjusted *R*

Table 3.1 also indicates unadjusted and adjusted values of the multiple correlation coefficient, R. The unadjusted values of R are simply the values of R for equations (3.1)–(3.3). The adjusted values of R are computed as partial R, where the control variables are U and I in the case of education; H, M, and I in the case of residence; and H, M, and U in the case of ethnicity. (Recall the earlier discussion of partial R in Chapter 2.)

Consider, for example, $R_{FMH \cdot UI}$ at the top of the rightmost column of Table 3.1. To calculate this partial R, we first regress F, H, and M on the control variables, which for this panel of the table are U and I:

$$\hat{F} = g + hU + jI \tag{3.8}$$

$$\hat{H} = k + pU + qI \tag{3.9}$$

$$\hat{M} = r + sU + tI \tag{3.10}$$

We then form the variables

$$F' = F - \hat{F} \tag{3.11}$$

$$H' = H - \hat{H} \tag{3.12}$$

$$M' = M - \hat{M} \tag{3.13}$$

Then we fit the regression

$$\hat{F}' = u + vH' + wM' \tag{3.14}$$

The multiple R for this equation is $R_{FMH \cdot UI}$.

When the main predictor variable is represented by a single dummy variable, partial R is the same as partial r, except that R is always considered positive. In basic MCA, R is used instead of r for the following reason: *When the values of the predictor variable represent categories, as in the case of basic MCA, it does not make sense to consider the correlation as a signed quantity, because the sign depends on which category is chosen as the reference category, which is arbitrary.* Because correlations are always considered positive in basic MCA, R is used throughout Table 3.1.

In the last line of Table 3.1, R for the full model with all predictor variables is simply R for equation (3.4). This R appears, for want of a better place, in the adjusted R column, even though it is not adjusted.

The notation used here for the unadjusted and adjusted multiple correlation coefficient, namely R, is not the notation conventionally used in MCA analyses. Conventionally η (eta) is used in place of unadjusted R, and β (beta) is used in place of adjusted (i.e., partial) R. We have opted for notation based on R because it seems preferable to reserve β to denote measures of effect (i.e., regression coefficients) and R to denote measures of goodness of fit. To use β in both senses is to invite confusion.

In ordinary multiple regression, R^2 (the coefficient of determination, representing the percentage of the variation in the response variable that is explained by the regression) is more commonly used than R itself to measure goodness of fit. But in MCA R (i.e., η or β instead of η^2 or β^2) is commonly used. The reason appears to be that partial R is often a very small number; use of R instead of R^2 conserves space in the MCA table by reducing the number of leading zeros that must be included in the tabulated values. In Table 3.1 we have changed the conventional notation from η and β to R, but we have followed MCA convention by retaining the unsquared multiple correlation coefficient.

3.1.4. A Numerical Example

When the formulae in Table 3.1 are applied to the Fiji Fertility Survey data, the result is Table 3.2. As a check on understanding the link between the underlying regressions and the MCA table, the reader

TABLE 3.2. MCA Table of the Effects of Education, Residence, and Ethnicity on Fertility[a]

Predictor Variable	n	Unadjusted Fertility	Unadjusted R	Adjusted Fertility	Adjusted R
Education			.12		.04
Low	573	6.72		6.48	
Medium	300	6.02		6.34	
High	132	5.76		6.04	
Residence			.12		.14
Urban	367	5.85		5.75	
Rural	638	6.69		6.74	
Ethnicity			.14		.15
Indian	554	6.81		6.84	
Fijian	451	5.86		5.82	
All predictor variables (full model)					.21

[a] Results in this table, which has the same form as Table 3.1, were generated by the MCA program in SPSS. The results may be checked by substituting appropriate combinations of ones, zeros, and mean values for H, M, U, and I in the following underlying regressions:

Unadjusted: Education: $\hat{F} = 6.716 - \underset{(.00)}{.696M} - \underset{(.00)}{.958H}$

Residence: $\hat{F} = 6.687 - \underset{(.00)}{.834U}$

Ethnicity: $\hat{F} = 5.856 + \underset{(.00)}{.955I}$

Adjusted: $\hat{F} = 6.283 - \underset{(.60)}{.140M} - \underset{(.19)}{.437H} - \underset{(.00)}{.991U} + \underset{(.00)}{1.016I}$

Numbers in parentheses under coefficients are p values; for example, a p value of .05 indicates a 5 percent level of significance. Note that the adjusted effects of education are not statistically significant at the 5 percent level. Mean values of the predictor variables are $\overline{M} = .299$, $\overline{H} = .131$, $\overline{U} = .365$, and $\overline{I} = .551$.

should recalculate the unadjusted and adjusted fertility values in Table 3.2 from the regressions reproduced in the footnote to the table. The results of this recalculation should agree with the numbers in Table 3.2 to within rounding errors in the last decimal place.

Recall from Chapter 2 that predicted values of the response variable are the same, regardless of which categories of the predictor variables are chosen as reference categories. Thus the numbers in Table 3.2 are the same, regardless of which categories of the predictor variables are chosen as reference categories.

Note from the full regression (labeled "adjusted" in the footnote to Tables 3.1 and 3.2) that the adjusted fertility effects of education are not statistically significant, because the p values for the coefficients of M and H exceed .05. The unadjusted effects of education, which are highly statistically significant, are explained away by residence and ethnicity.

Residence and ethnicity have mostly independent effects on fertility, insofar as the unadjusted fertility values by residence and ethnicity are affected very little by the introduction of controls.

The R values are very low in Table 3.2, indicating that the predictor variables do not explain much of the variation of fertility.

Table 3.2 was produced by the MCA program in SPSS. SPSS is the only major statistical package with an MCA program that produces MCA tables directly. When using other packages, one must calculate MCA tables from the underlying multiple regressions.

3.2. THE MCA TABLE IN DEVIATION FORM

An alternative form of the MCA table presents unadjusted and adjusted fertility values as *deviations from the grand mean*. In our example using Fiji data, the grand mean is $\bar{F}$, the mean of F in the entire sample of ever-married women aged 40–49.

3.2.1. First Approach to Table Set-up

The simplest way to convert Table 3.1 to deviation form is to subtract $\bar{F}$ from each cell entry in the unadjusted and adjusted fertility columns. For example, the formula for the unadjusted fertility value for medium education, expressed in deviation form, is $a' + b' - \bar{F}$.

Effects, again represented by differences between fertility values for pairs of categories of the predictor variable, are unaffected by subtracting $\bar{F}$ from each cell entry, because the $\bar{F}$ terms cancel when taking differences.

3.2.2. Second Approach to Table Set-up

The second approach to setting up the MCA table in deviation form is shown in Table 3.3. The second approach is equivalent to the first approach of subtracting $\bar{F}$ from each cell entry, insofar as the two approaches yield the same numerical results.

TABLE 3.3. Alternative Set-up for the MCA Table of the Effects of Education, Residence, and Ethnicity on Fertility, with Fertility Effects Reexpressed as Deviations from the Grand Mean[a]

Predictor Variable	n	Unadjusted Fertility	Unadjusted R	Adjusted Fertility	Adjusted R
Education			R_{FMH}		$R_{FMH \cdot UI}$
Low	n_L	$-b'\overline{M} - c'\overline{H}$		$-b\overline{M} - c\overline{H}$	
Medium	n_M	$b'(1 - \overline{M}) - c'\overline{H}$		$b(1 - \overline{M}) - c\overline{H}$	
High	n_H	$-b'\overline{M} + c'(1 - \overline{H})$		$-b\overline{M} + c(1 - \overline{H})$	
Residence			R_{FU}		$R_{FU \cdot MHI}$
Urban	n_U	$d''(1 - \overline{U})$		$d(1 - \overline{U})$	
Rural	n_R	$-d''\overline{U}$		$-d\overline{U}$	
Ethnicity			R_{FI}		$R_{FI \cdot MHU}$
Indian	n_I	$e'''(1 - \overline{I})$		$e(1 - \overline{I})$	
Fijian	n_F	$-e'''\overline{I}$		$-e\overline{I}$	
All predictor variables (full model)					R

[a]The underlying regressions are:

Unadjusted: Education: $\hat{F} - \overline{F} = b'(M - \overline{M}) + c'(H - \overline{H})$

Residence: $\hat{F} - \overline{F} = d''(U - \overline{U})$

Ethnicity: $\hat{F} - \overline{F} = e'''(I - \overline{I})$

Adjusted: $\hat{F} - \overline{F} = b(M - \overline{M}) + c(H - \overline{H}) + d(U - \overline{U}) + e(I - \overline{I})$

In Table 3.3, the formulae in the unadjusted and adjusted fertility columns are derived from a reformulation of the underlying regressions. Consider first the underlying regression for the unadjusted values of F by education, from Table 3.1:

$$\hat{F} = a' + b'M + c'H \quad \text{(3.1 repeated)}$$

Because the regression plane passes through the point $(\overline{M}, \overline{H}, \overline{F})$, we have from (3.1) that

$$\overline{F} = a' + b'\overline{M} + c'\overline{H} \quad (3.15)$$

Subtracting (3.15) from (3.1), we obtain

$$\hat{F} - \overline{F} = b'(M - \overline{M}) + c'(H - \overline{H}) \quad (3.16)$$

This equation directly expresses the predicted values of F as deviations from the grand mean. *Note that* (3.1) *and* (3.16) *are alternative forms of the same underlying regression.*

Similarly, the underlying regression for the unadjusted fertility values by residence can be expressed in deviation form as

$$\hat{F} - \bar{F} = d''(U - \bar{U}) \tag{3.17}$$

and the underlying regression for the unadjusted fertility values by ethnicity can be expressed in deviation form as

$$\hat{F} - \bar{F} = e'''(I - \bar{I}) \tag{3.18}$$

The formulae for unadjusted fertility values in deviation form in Table 3.3 are derived by setting the predictor variables to appropriate combinations of ones and zeros in (3.16)–(3.18), just as before. For example, the entry for medium education, $b'(1 - \bar{M}) - c'\bar{H}$, is obtained by setting $M = 1$ and $H = 0$ in (3.16). *When numerical values are plugged in,* $b'(1 - \bar{M}) - c'\bar{H}$ *in Table* 3.3 *yields the same numerical result as* $a' + b' - \bar{F}$*, which is what one gets by subtracting* $\bar{F}$ *from the medium education entry,* $a' + b'$*, in Table* 3.1. Similarly, the unadjusted fertility entry for rural, $-d''\bar{U}$, in Table 3.3 is obtained by setting $U = 0$ in (3.17). When numerical values are plugged in, $-d''\bar{U}$ in Table 3.3 yields the same numerical result as $a'' - \bar{F}$, which is what one gets by subtracting $\bar{F}$ from the rural entry, a'', in Table 3.1.

Turning now to the underlying regression for the adjusted fertility values, we have

$$\hat{F} = a + bM + cH + dU + eI \qquad \text{(3.4 repeated)}$$

$$\bar{F} = a + b\bar{M} + c\bar{H} + d\bar{U} + e\bar{I} \tag{3.19}$$

$$\hat{F} - \bar{F} = b(M - \bar{M}) + c(H - \bar{H}) + d(U - \bar{U}) + e(I - \bar{I}) \tag{3.20}$$

The formulae for adjusted fertility values in Table 3.3 are expressed in deviation form by setting the predictor variables to appropriate combinations of ones, zeros, and mean values in this equation. For example, the entry for medium education, $b(1 - \bar{M}) - c\bar{H}$, is obtained by setting $M = 1$, $H = 0$, $U = \bar{U}$, and $I = \bar{I}$ in (3.20). Note how the last two terms in (3.20) disappear when $U = \bar{U}$ and $I = \bar{I}$. When numerical values are plugged in, $b(1 - \bar{M}) - c\bar{H}$ in Table 3.3 yields the same numerical results as $a + b + d\bar{U} + e\bar{I} - \bar{F}$, which is what one gets by subtracting $\bar{F}$ from the medium education entry, $a + b + d\bar{U} + e\bar{I}$, in Table 3.1. Similarly, the adjusted fertility deviation for rural, $-d\bar{U}$, in Table 3.3 is obtained by setting $U = 0$, $M = \bar{M}$, $H = \bar{H}$, and $I = \bar{I}$ in (3.20). When numerical values are plugged in, $-d\bar{U}$ in Table 3.3

yields the same numerical result as $a + b\overline{M} + c\overline{H} + e\bar{I} - \overline{F}$, which is what one gets by subtracting $\overline{F}$ from the rural entry, $a + b\overline{M} + c\overline{H} + e\bar{I}$, in Table 3.1.

Note the parallel algebraic form of the formulae for unadjusted and adjusted fertility values expressed in deviation form in Table 3.3. The formulae for adjusted values of fertility by education, residence, and ethnicity are compact and would remain unchanged even if the table were lengthened by adding more predictor variables. This is so because terms containing the control variables drop out of these formulae when the control variables are set at their mean values.

3.2.3. A Numerical Example

When the formulae in Table 3.3 are applied to the Fiji Fertility Survey data, the result is Table 3.4. The underlying regressions given in the footnote to this table can be used to derive the unadjusted and

TABLE 3.4. MCA Table of the Effects of Education, Residence, and Ethnicity on Fertility, with Fertility Effects Expressed as Deviations from the Grand Mean[a]

Predictor Variable	n	Unadjusted Fertility	Unadjusted R	Adjusted Fertility	Adjusted R
Education			.12		.04
Low	573	.33		.10	
Medium	300	−.36		−.04	
High	132	−.62		−.34	
Residence			.12		.14
Urban	367	−.53		−.63	
Rural	638	.30		.36	
Ethnicity			.14		.15
Indian	554	.43		.46	
Fijian	451	−.53		−.56	
All predictor variables (full model)					.21

[a]The grand mean is 6.38. The fertility values in this table may be generated either by subtracting 6.38 from each unadjusted or adjusted fertility value in Table 3.2, or by plugging in appropriate combinations of ones, zeros, and mean values in the following regressions:

Unadjusted: Education: $\hat{F} - \overline{F} = -.696(M - .299) - .958(H - .131)$

Residence: $\hat{F} - \overline{F} = -.834(U - .365)$

Ethnicity: $\hat{F} - \overline{F} = .955(I - .551)$

Adjusted: $\hat{F} - \overline{F} = -.140(M - .299) - .437(H - .131) - .991(U - .365) + 1.016(I - .551)$

adjusted fertility values in deviation form. Alternatively, the fertility deviations from the grand mean in Table 3.4 can be obtained by subtracting $\bar{F} = 6.38$ from the corresponding fertility values in Table 3.2. (Because of rounding errors, the results of these two calculations may differ in the last decimal place.)

Table 3.4, in deviation form, is more abstract and difficult to read than Table 3.2, in absolute form, especially for the intelligent layman or policymaker to whom results are sometimes directed. Therefore, most authors prefer to present the MCA table in absolute form instead of deviation form. In the remainder of this chapter, MCA tables are presented in absolute form only.

3.3. MCA WITH INTERACTIONS

Suppose that the response variable is once again fertility, and that we have three predictor variables: education (low, medium, and high), residence (urban, rural), and ethnicity (Indian, Fijian). Suppose also that there is interaction between residence and ethnicity.

3.3.1. Table Set-up

The underlying regression for unadjusted fertility values for categories of education is

$$\hat{F} = a' + b'M + c'H \tag{3.21}$$

The underlying regression for unadjusted fertility values for cross-classified categories of residence and ethnicity (the two interacting variables) is

$$\hat{F} = a'' + d''U + e''I + f''UI \tag{3.22}$$

The underlying regression for adjusted fertility values is

$$\hat{F} = a + bM + cH + dU + eI + fUI \tag{3.23}$$

where F, M, H, U, and I are defined as before.

Because the effect of U on F depends on the level of I, and the effect of I on F depends on the level of U, we cannot present the effects of U and I separately. *When there is interaction between two predictor variables, we must cross-classify the categories of these two variables in the MCA table, and we use an interactive model for calculating unadjusted and adjusted fertility values for the cross-classified categories*, as shown in equations (3.22) and (3.23) and in Table 3.5.

In Table 3.5, the formulae for unadjusted and adjusted fertility values are derived by setting the predictor variables to appropriate combinations of ones, zeros, and mean values in (3.21)–(3.23).

The formulae for unadjusted fertility values for categories of education are the same as those in Table 3.1. These entries are unchanged because education is not one of the interacting variables. On the other hand, the formulae for unadjusted fertility values for categories of residence cross-classified by ethnicity are obtained by setting (U, I) alternatively to $(1, 1)$, $(1, 0)$, $(0, 1)$, and $(0, 0)$ in (3.22). In each of these four cases, the value of UI is determined from the assigned values of U and I.

TABLE 3.5. Set-up for MCA Table of the Effects of Education, Residence, and Ethnicity on Fertility, with Interaction Between Residence and Ethnicity[a]

Predictor Variable	n	Unadjusted Fertility	Unadjusted R	Adjusted Fertility	Adjusted R
Education			R_{FMH}		$R_{FMH \cdot UI(UI)}$
Low	n_L	a'		$a + d\bar{U} + e\bar{I} + f\overline{UI}$	
Medium	n_M	$a' + b'$		$a + b + d\bar{U} + e\bar{I} + f\overline{UI}$	
High	n_H	$a' + c'$		$a + c + d\bar{U} + e\bar{I} + f\overline{UI}$	
Residence and ethnicity			$R_{FUI(UI)}$		$R_{FUI(UI) \cdot MH}$
Urban Indian	n_{UI}	$a'' + d'' + e'' + f''$		$a + d + e + f + b\bar{M} + c\bar{H}$	
Urban Fijian	n_{UF}	$a'' + d''$		$a + d + b\bar{M} + c\bar{H}$	
Rural Indian	n_{RI}	$a'' + e''$		$a + e + b\bar{M} + c\bar{H}$	
Rural Fijian	n_{RF}	a''		$a + b\bar{M} + c\bar{H}$	
All predictor variables (full model)					R

[a]The underlying regressions are:

Unadjusted: Education: $\hat{F} = a' + b'M + c'H$

Residence × ethnicity: $\hat{F} = a'' + d''U + e''I + f''UI$

Adjusted: $\hat{F} = a + bM + cH + dU + eI + fUI$

When UI is found in parentheses in the subscript of R, it indicates that a UI term is included in the underlying regression.

The formulae for adjusted fertility values by education are obtained by setting (M, H) alternatively to (0, 0), (1, 0), and (0, 1) in the equation

$$\hat{F} = a + bM + cH + d\overline{U} + e\overline{I} + f\overline{UI} \tag{3.24}$$

where (3.24) is obtained by setting $U = \overline{U}$, $I = \overline{I}$, and $UI = \overline{UI}$ in (3.23). (Because UI is treated as a separate variable, the mean is calculated as the mean of UI instead of the mean of U times the mean of I. Note that the mean of UI, namely $\overline{UI}$, is the sample proportion who are both urban and Indian.[2]) The formulae for adjusted fertility values by categories of residence cross-classified by ethnicity are obtained by setting (U, I) alternatively to (1, 1), (1, 0), (0, 1), and (0, 0) in the equation

$$\hat{F} = a + dU + eI + fUI + b\overline{M} + c\overline{H} \tag{3.25}$$

where (3.25) is obtained by setting $M = \overline{M}$ and $H = \overline{H}$ in (3.23).

Partial R is calculated in the usual way. When UI is found in parentheses in the subscript of R in Table 3.5, it simply indicates that a UI term is included in the underlying regression.

Unfortunately, the SPSS MCA program cannot handle interactions very well. One can ask for no interactions or all interactions of a given order, such as all two-way interactions, but not selected two-way interactions. Therefore, when the model is interactive, one often must compute the MCA table from the underlying multiple regressions. Partial R must be computed separately, by the approach explained earlier.

3.3.2. A Numerical Example

When the formulae in Table 3.5 are applied to the Fiji Fertility Survey data, the result is Table 3.6. As a check on understanding, the reader

[2]This introduces an inconsistency, because in general the mean of UI does not equal the mean of U times the mean of I. But this inconsistency is of no consequence, because the levels at which the control variables are held constant are a matter of choice and therefore somewhat arbitrary, as explained earlier.

TABLE 3.6. MCA Table of the Effects of Education, Residence, and Ethnicity on Fertility, with Interaction Between Race and Ethnicity[a]

Predictor Variable	n	Unadjusted Fertility	Unadjusted R	Adjusted Fertility	Adjusted R
Education			.12		.04
Low	573	6.72		6.49	
Medium	300	6.02		6.34	
High	132	5.76		6.03	
Residence and ethnicity			.21		.18
Urban Indian	243	6.33		6.31	
Urban Fijian	124	4.92		4.99	
Rural Indian	311	7.19		7.11	
Rural Fijian	327	6.21		6.27	
All predictor variables (full model)					.21

[a]The underlying regressions are:

Unadjusted: Education: $\hat{F} = 6.716 - \underset{(.00)}{.696M} - \underset{(.00)}{.958H}$

Residence × ethnicity:

$$\hat{F} = 6.211 - \underset{(.00)}{1.292U} + \underset{(.00)}{.975I} + \underset{(.32)}{.434UI}$$

Adjusted: $\hat{F} = 6.371 - \underset{(.57)}{.151M} - \underset{(.17)}{.458H} - \underset{(.00)}{1.273U} + \underset{(.00)}{.849I} + \underset{(.28)}{.472UI}$

Numbers in parentheses are p values. Mean values of the predictor variables are $\overline{M} = .299$, $\overline{H} = .131$, $\overline{U} = .365$, $\overline{I} = .551$, and $\overline{UI} = .242$. We had to generate the MCA table from the underlying multiple regressions and partial correlations, because the MCA program in SPSS was not flexible enough to handle this particular specification of interaction. The adjusted effects of education on fertility are not statistically significant at the 5 percent level. This also applies to the interaction effects, whether unadjusted or adjusted.

may use the underlying regressions given in the footnote to the table to replicate the numerical entries in the unadjusted and adjusted fertility columns.

3.4. MCA WITH ADDITIONAL QUANTITATIVE CONTROL VARIABLES

Suppose that (1) the response variable is fertility, (2) the predictor variables are education (low, medium, high), residence (urban, rural), and ethnicity (Indian, Fijian), and (3) the additional quantitative control variable is age at first marriage, denoted by A.

3.4.1. Table Set-up

The underlying regressions without controls are

$$\hat{F} = a' + b'M + c'H \tag{3.26}$$

$$\hat{F} = a'' + d''U \tag{3.27}$$

$$\hat{F} = a'' + e''I \tag{3.28}$$

TABLE 3.7. Set-up for MCA Table of the Effects of Education, Residence, and Ethnicity on Fertility, with Age at First Marriage (a Quantitative Variable) Statistically Controlled[a]

Predictor Variable	n	Unadjusted Fertility	Unadjusted R	Adjusted Fertility	Adjusted R
Education			R_{FMH}		$R_{FMH \cdot UIA}$
Low	n_L	a'		$a + d\bar{U} + e\bar{I} + f\bar{A}$	
Medium	n_M	$a' + b'$		$a + b + d\bar{U} + e\bar{I} + f\bar{A}$	
High	n_H	$a' + c'$		$a + c + d\bar{U} + e\bar{I} + f\bar{A}$	
Residence			R_{FU}		$R_{FU \cdot MHIA}$
Urban	n_U	$a'' + d''$		$a + d + b\bar{M} + c\bar{H} + e\bar{I} + f\bar{A}$	
Rural	n_R	a''		$a + b\bar{M} + c\bar{H} + e\bar{I} + f\bar{A}$	
Ethnicity			R_{FI}		$R_{FI \cdot MHUA}$
Indian	n_I	$a''' + e'''$		$a + e + b\bar{M} + c\bar{H} + d\bar{U} + f\bar{A}$	
Fijian	n_F	a'''		$a + b\bar{M} + c\bar{H} + d\bar{U} + f\bar{A}$	
All predictor variables (full model)					R

[a] The underlying regressions are:

Unadjusted: Education: $\hat{F} = a' + b'M + c'H$

Residence: $\hat{F} = a'' + d''U$

Ethnicity: $\hat{F} = a''' + e'''I$

Adjusted: $\hat{F} = a + bM + cH + dU + eI + fA$

and the underlying regression with controls is

$$\hat{F} = a + bM + cH + dU + eI + fA \tag{3.29}$$

where F, M, H, U, and I are defined as before.

The set-up for the MCA table is shown in Table 3.7.

3.4.2. A Numerical Example

Illustrative results for Fiji are shown in Table 3.8. Tables 3.7 and 3.8 have the same format as Tables 3.1 and 3.2 but contain an additional

TABLE 3.8. MCA Table of the Effects of Education, Residence, and Ethnicity on Fertility, with Age at First Marriage (a Quantitative Variable) Statistically Controlled[a]

Predictor Variable	n	Unadjusted		Adjusted	
		Fertility	R	Fertility	R
Education			.12		.01
Low	573	6.72		6.35	
Medium	300	6.02		6.45	
High	132	5.76		6.35	
Residence			.12		.14
Urban	367	5.85		5.77	
Rural	638	6.69		6.73	
Ethnicity					
Indian	554	6.81	.14	6.56	.06
Fijian	451	5.86		6.15	
All predictor variables (full model)					.33

[a]The underlying regressions are:

Unadjusted:

Education: $\hat{F} = 6.716 - \underset{(.00)}{.696M} - \underset{(.00)}{.958H}$

Residence: $\hat{F} = 6.687 - \underset{(.00)}{.834U}$

Ethnicity: $\hat{F} = 5.856 + \underset{(.00)}{.955I}$

Adjusted: $\hat{F} = 10.253 + \underset{(.71)}{.097M} - \underset{(.99)}{.0006H} - \underset{(.00)}{.967U} + \underset{(.10)}{.409I} - \underset{(.00)}{.212A}$

Numbers in parentheses are p values. Means of the predictor variables are $\overline{M} = .299$, $\overline{H} = .131$, $\overline{U} = .365$, $\overline{I} = .551$, and $\overline{A} = 17.789$. The adjusted effects of education and the adjusted effect of ethnicity are not statistically significant at the 5 percent level.

control, namely age at first marriage, which affects the adjusted fertility values and the adjusted values of R. The unadjusted parts of Tables 3.7 and 3.8 are the same as the unadjusted parts of Tables 3.1 and 3.2.

3.5. EXPRESSING RESULTS FROM ORDINARY MULTIPLE REGRESSION IN AN MCA FORMAT (ALL VARIABLES QUANTITATIVE)

Suppose that we wish to assess the effect of education and age at first marriage on fertility, where the variables (all quantitative) are now defined as

F: fertility (number of children ever born)
E: number of completed years of education
A: age at first marriage

We can express results in an MCA format by displaying values of F for selected values of the predictor variables. For tabulation purposes, the selected values of the predictor variables play the role of categories. Otherwise the mathematics is the same as before.

3.5.1. Table Set-up

In Table 3.9, the selected values of E are 0, 6, 12, and 16, and the selected values of A are 15, 20, 25, and 30. The underlying regressions for the unadjusted fertility values are

$$\hat{F} = a' + b'E \tag{3.30}$$

$$\hat{F} = a'' + c''A \tag{3.31}$$

and the underlying regression for the adjusted fertility values is

$$\hat{F} = a + bE + cA \tag{3.32}$$

The formulae for unadjusted fertility values in Table 3.9 are obtained by setting the predictor values at their selected values in (3.30) and (3.31). The formulae for adjusted fertility values in Table 3.9 are obtained from (3.32) by setting the predictor variable of interest at its

TABLE 3.9. Set-up for MCA Table of the Effects of Education and Age at First Marriage on Fertility (All Variables Quantitative)[a]

Predictor Variable	Unadjusted		Adjusted	
	Fertility	r	Fertility	r or R
Education		r_{FE}		$r_{FE \cdot A}$
0	a'		$a + c\bar{A}$	
6	$a' + 6b'$		$a + 6b + c\bar{A}$	
12	$a' + 12b'$		$a + 12b + c\bar{A}$	
16	$a' + 16b'$		$a + 16b + c\bar{A}$	
Age at first marriage		r_{FA}		$r_{FA \cdot E}$
15	$a'' + 15c''$		$a + 15c + b\bar{E}$	
20	$a'' + 20c''$		$a + 20c + b\bar{E}$	
25	$a'' + 25c''$		$a + 25c + b\bar{E}$	
30	$a'' + 30c''$		$a + 30c + b\bar{E}$	
All predictor variables (full model)				R

[a]The underlying regressions are:

Unadjusted: Education: $\hat{F} = a' + b'E$

Age at first marriage: $\hat{F} = a'' + c''A$

Adjusted: $\hat{F} = a + bE + cA$

selected values and setting the other variable, which plays the role of control variable, at its mean value in the entire sample.

For measure of goodness of fit, we now use partial r instead of partial R, because the correlations are all bivariate and because, with quantitative variables instead of categorical variables, it now makes sense to attach a positive or negative sign to the correlation.

3.5.2. A Numerical Example

When the formulae in Table 3.9 are applied to the Fiji Fertility Survey data, the result is Table 3.10.

3.5.3. Why Present Ordinary Regression Results in an MCA Format?

The mode of presentation in Tables 3.9 and 3.10 is useful for presenting results to a broader audience, because the MCA table is less abstract and is easier to understand than a table of regression coeffi-

TABLE 3.10. MCA Table of the Effects of Education and Age at First Marriage on Fertility (All Variables Quantitative)[a]

Predictor Variable	Unadjusted		Adjusted	
	Fertility	r	Fertility	r or R
Education		−.14		−.03
0	6.89		6.50	
6	6.05		6.30	
12	5.22		6.10	
16	4.66		5.97	
Age at first marriage		−.30		−.27
15	7.01		6.98	
20	5.88		5.90	
25	4.75		4.82	
30	3.63		3.74	
All predictor variables (full model)				.30

[a]The underlying regressions are:

Unadjusted: Education: $\hat{F} = 6.885 - \underset{(.00)}{.139E}$

Age at marriage:

$$\hat{F} = 10.380 - \underset{(.00)}{.225A}$$

Adjusted: $\hat{F} = 10.343 - \underset{(.32)}{.033E} - \underset{(.00)}{.216A}$

Numbers in parentheses are p values. Mean values of the predictor variables are $\overline{E} = 3.622$ and $\overline{A} = 17.789$. Regressions were generated by the multiple regression program, and partial correlations were generated by the partial correlation program in SAS. The adjusted effect of education is not statistically significant at the 5 percent level.

cients. Actually, this is a general point that can be made about all MCA tables.

3.6. PRESENTING MCA RESULTS GRAPHICALLY

Because the adjusted MCA table comprises a set of smaller bivariate tables for which the response variable is quantitative, each of these smaller tables can be graphed.

When the response variable is quantitative and the predictor variable is a quantitative variable grouped into categories, the adjusted

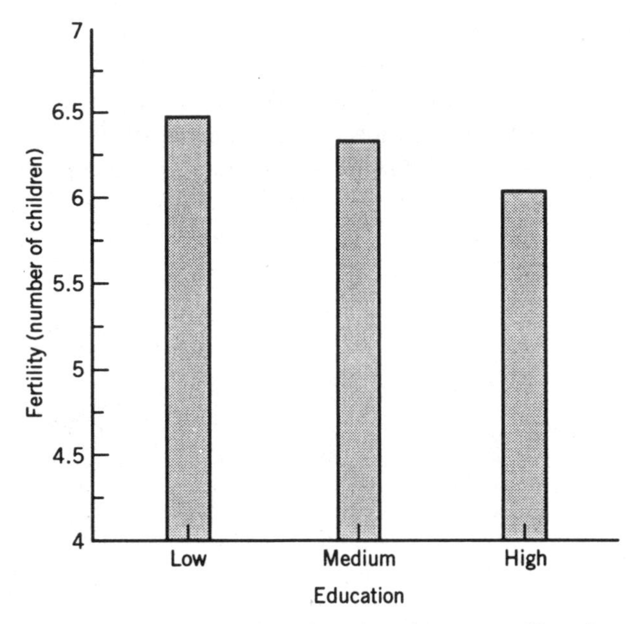

Figure 3.1. Adjusted values of fertility by education, controlling for residence and ethnicity. **Note:** Based on adjusted values of fertility by education in Table 3.2.

values of the response variable can be represented by a piecewise-linear curve of the sort discussed in Figure 2.6b in Chapter 2. When the predictor variable is purely qualitative, adjusted values of the response variable can be graphed as a bar chart, with a separate bar for each category of the predictor variable. As an example, Figure 3.1 shows a bar chart of fertility against education (low, medium, high), with residence and ethnicity controlled, based on Table 3.2.

In the case where both the response variable and the predictor variable are quantitative, the adjusted values of the response variable can be plotted against the predictor variable simply as a straight line. In the example of the previous section, in which adjusted fertility values were derived from equation (3.32),

$$\hat{F} = a + bE + cA \qquad \text{(3.32 repeated)}$$

the plot of adjusted fertility by, say, education, is simply

$$\hat{F} = (a + c\bar{A}) + bE \qquad (3.33)$$

which can be written $\hat{F} = a^* + bE$, where the constant term a^* denotes $a + c\bar{A}$.

3.7. FURTHER READING

Multiple Classification Analysis, a book by Andrews, Morgan, and Sonquist (1969), outlines the method and a computer program, but the presentation is difficult. Chapter 4 of *The American Occupational Structure*, by Blau and Duncan (1967), provides a more readable exposition.

4

PATH ANALYSIS

In multiple regression, each predictor variable has a direct effect on the response variable. What if the predictor variable affects the response variable not only directly but also indirectly through one or more intervening variables? Path analysis is a technique for analyzing this kind of causal relationship. It builds on ordinary multiple regression.

4.1. PATH DIAGRAMS AND PATH COEFFICIENTS

The simplest path model is in the form of a single regression equation. As an example, let us consider the model

$$\hat{F} = a + bU \tag{4.1}$$

fitted to women aged 35–49 in the 1974 Fiji Fertility Survey, where, for each woman,

F = number of children ever born
U = 1 if urban, 0 otherwise

Suppose that we have a theory that says that U is a cause of variation in F. To view (4.1) in the context of path analysis, the first step is to portray it in a *path diagram*, shown in Figure 4.1a. In path diagrams, arrows between variables indicate the direction of causa-

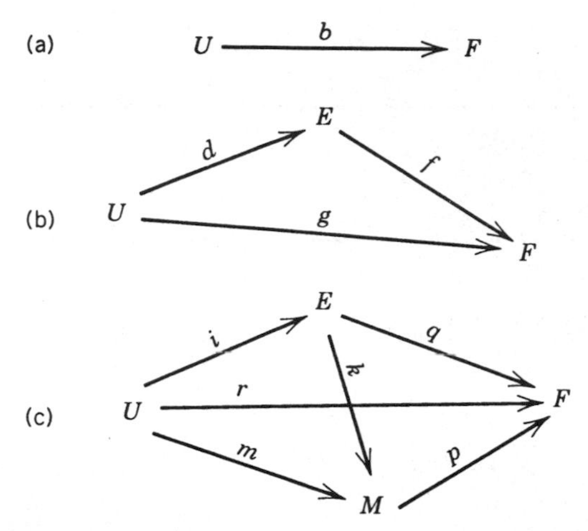

Figure 4.1. Path diagrams for modeling the effects of residence, education, and age at first marriage on fertility. **Notes**: The equations for the three models are

Model (a)

$$F = a + bU$$

Model (b)

$$E = c + dU$$
$$F = e + fE + gU$$

Model (c)

$$E = h + iU$$
$$M = j + kE + mU$$
$$F = n + pM + qE + rU$$

tion, and numbers written on the arrows represent effects in the form of regression coefficients. In path analysis these regression coefficients are called *path coefficients*. The causal ordering, as specified by the path diagram, is derived from theory, not from the data themselves.

A somewhat more adequate model might include a third variable, E (number of completed years of education), where we theorize that the three variables are causally related as in Figure 4.1b. Because U has no arrows coming toward it, it is not determined by other variables in the model and is therefore called an *exogenous variable*. In contrast, E and F are *endogenous variables*, because they have arrows coming toward them and are therefore determined, at least in part, by other variables in the model.

In Figure 4.1b, there are no feedback loops; that is, we cannot start from a variable, follow the arrows, and get back to the same variable.

An example of a feedback loop would be the addition of a return arrow between, say, U and E, yielding the *direct feedback loop* $U \rightleftarrows E$. Another example would be reversal of the direction of the arrow between U and F, which would establish an *indirect feedback loop*, $U \rightarrow E \rightarrow F \rightarrow U$.

A model without feedback loops and with uncorrelated errors (between equations as well as within equations) is called a *recursive model*. In this chapter we shall consider only recursive models, because *nonrecursive models*, containing feedback loops, are more difficult to estimate. The simple recursive models considered here belong to a broader class of models called *structural equation models*, which include nonrecursive models.

In the model in Figure 4.1b, *a regression equation can be written for each endogenous variable*. In each such equation, the endogenous variable is the response variable, and other variables with arrows coming directly to the response variable are the predictor variables. The estimated model in Figure 4.1b has two endogenous variables, E and F, and is accordingly represented by two equations:

$$\hat{E} = c + dU \tag{4.2}$$

$$\hat{F} = e + fE + gU \tag{4.3}$$

Because (4.2) and (4.3), taken together, represent a single model, they should be estimated simultaneously. If the model is recursive, as this one is, and if the errors in the underlying population regressions are statistically independent, as we assume they are, then it can be shown that simultaneous estimation of (4.2) and (4.3) is equivalent to estimating (4.2) and (4.3) separately by ordinary least-squares regression. *This means that we can treat path analysis as an extension of ordinary multiple regression*.

In the estimated model represented by equations (4.2) and (4.3), F is the ultimate response variable and U is the exogenous variable. By substituting (4.2) into (4.3), we can represent F solely in terms of the exogenous variable U. The resulting equation, called the *reduced form* of the model, can be written

$$\hat{F} = (e + cf) + (g + df)U \tag{4.4}$$

In (4.4), the coefficient g is called the *direct effect* of U on F, and the product of coefficients df is called the *indirect effect* of U on F. In Figure 4.1b, the direct effect, g, is written on the arrow that goes directly from U to F; this arrow is called the *direct path* from U to F. The indirect effect, df, is the product of the coefficient d over the arrow from U to E and the coefficient f over the arrow from E to F. This path, because it involves an intervening variable E, is called an *indirect path* from U to F. We shall denote the direct path as $U \rightarrow F$ and the indirect path as $U \rightarrow E \rightarrow F$. Direct paths are sometimes called *simple paths*, and indirect paths are sometimes called *compound paths*.

Intuitively, it makes sense that the indirect effect is df. If a one-unit increase in U augments E by d units, and a one-unit increase in E augments F by f units, then, over the entire path, a one-unit increase in U must augment F by df units.

When we compare equations (4.1) and (4.4), we see that they have the same form. Thus $a = e + cf$ and $b = g + df$. The coefficient b is called the *total effect* of U on F. *The total effect equals the sum of the direct effect and the indirect effect.*

Notice that the addition of the intervening variable E to the model in Figure 4.1*b does not invalidate the simpler model in Figure* 4.1*a*. The total effect of U on F is the same in either case. The addition of E to the model simply decomposes the total effect into a direct effect and an indirect effect.

Let us further complicate the model by adding age at first marriage, M, as a second intervening variable between U and F. In the Fiji Fertility Survey, very few women were still in school by the time they got married, so that we may reasonably assume that E is causally prior to M. The time ordering of the variables, which we take as a causal ordering, is U, E, M, F. In the path diagram in Figure 4.1c, the variables are arrayed from left to right to reflect this time ordering.

The estimated model equations are

$$\hat{E} = h + iU \tag{4.5}$$

$$\hat{M} = j + kE + mU \tag{4.6}$$

$$\hat{F} = n + pM + qE + rU \tag{4.7}$$

The reduced from of the model is obtained by substituting (4.5) and (4.6) into (4.7). After some algebraic manipulation, these substitutions yield

$$\hat{F} = (n + jp + hq + hkp) + (r + iq + mp + ikp)U \tag{4.8}$$

This equation has the same mathematical form as equation (4.1). Therefore $a = n + jp + hq + hkp$ and $b = r + iq + mp + ikp$. The total effect of U on F is b, the same as before. The total effect is decomposed into a direct effect r (corresponding to the direct path $U \rightarrow F$) and indirect effects iq (corresponding to the indirect path $U \rightarrow E \rightarrow F$), mp (corresponding to the indirect path $U \rightarrow M \rightarrow F$), and ikp (corresponding to the indirect path $U \rightarrow E \rightarrow M \rightarrow F$). *It is evident that one can obtain the total effect in either of two ways*: (1) *by computing the coefficient of U in the reduced form of the model,* $\hat{F} = a + bU$, *or* (2) *by tracing out all possible paths from U to F, computing the product of coefficients along each path, and summing the products over all possible paths.*

In the same way, we can calculate direct, indirect, and total effects of E on F. To derive these effects, we eliminate the intervening variable M by substituting (4.6) into (4.7), yielding the *partially reduced form*

$$\hat{F} = (n + jp) + (q + kp)E + (r + mp)U \tag{4.9}$$

The coefficient q is the direct effect of E on F via the path $E \rightarrow F$; the product kp is the indirect effect of E on F via the path $E \rightarrow M \rightarrow F$; and the sum $q + kp$ is the total effect of E on F. *In the partially reduced form in* (4.9), *the direct, indirect, and total effects of E on F all incorporate a control for the causally prior variable U because U also appears in* (4.9). (Because of this control, some authors do not call $q + kp$ a total effect.) In (4.9), the coefficient of U, $r + mp$, is not of much interest because it is not the total effect of U on F, insofar as it omits indirect effects of U on F via E.

Because equation (4.9) has the same form as equation (4.3), it is evident that the total effect of E on F in model (c) is the same as the direct effect of E on F in model (b) in Figure 4.1. In model (b), the direct effect of E on F is the same as the total effect of E on F. *Thus the total effect of E on F is the same in models* (*b*) *and* (*c*).

When the equations for models (a), (b), and (c) in Figure 4.1 are fitted to the data, the result is

Model (a)

$$\hat{F} = 6.1887 - .8209U \tag{4.10}$$

Model (b)

$$E = 3.1769 + .6432U \tag{4.11}$$

$$\hat{F} = 6.7977 - .1616E - .7169U \tag{4.12}$$

Model (c)

$$\hat{E} = 3.1769 + .6432U \tag{4.13}$$

$$\hat{M} = 16.2150 + .4716E - .5366U \tag{4.14}$$

$$\hat{F} = 10.2533 - .0611E - .2132M - .8313U \tag{4.15}$$

Figure 4.2 shows path diagrams for the fitted models, and Table 4.1 summarizes direct, indirect, and total effects of U on F and E on F. (In both the figure and the table, effects are rounded to two decimal places.) As expected, the total effects of U on F are the same for all three models, and the total effects of E on F are the same for models (b) and (c), which are the only models that include E. Table 4.1 shows that the effect of residence on fertility is almost entirely direct, whereas the effect of education on fertility is mostly indirect through age at marriage.

Equations (4.10)–(4.15) and Figure 4.2 provide no indication of statistical significance of coefficients. Throughout most of the rest of this chapter, we shall not be concerned about whether the coefficients are statistically significant, although this would be of interest in a real research project.

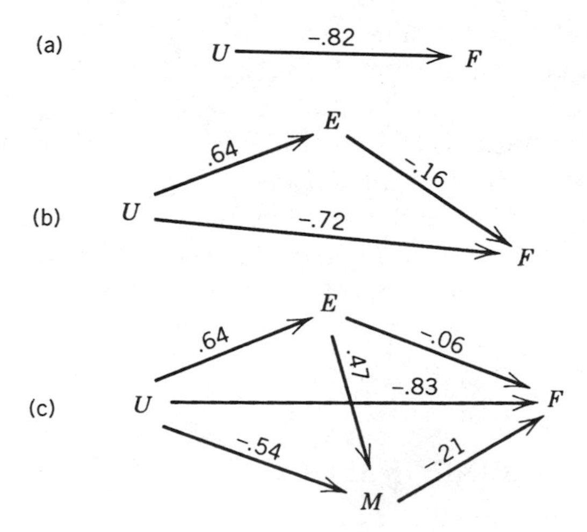

Figure 4.2. Fitted path models of the effects of residence, education, and age at first marriage on fertility: women aged 35–49, 1974 Fiji Fertility Survey. **Note**: The models in this figure are fitted versions of the models in Figure 4.1.

TABLE 4.1. Direct, Indirect, and Total Effects of Residence and Education on Fertility: Ever-Married Women Ages 35–49, 1974 Fiji Fertility Survey[a]

		Model					
		(a)		(b)		(c)	
Variable	Type of Effect	Formula	Numerical Value	Formula	Numerical Value	Formula	Numerical Value
Residence	Direct	b	−.82	g	−.72	r	−.83
	Indirect						
	$U \rightarrow E \rightarrow F$	—	—	df	−.10	iq	−.04
	$U \rightarrow M \rightarrow F$	—	—	—	—	mp	.11
	$U \rightarrow E \rightarrow M \rightarrow F$	—	—	—	—	ikp	−.06
	Total	b	−.82	b	−.82	b	−.82
Education	Direct	—	—	f	−.16	q	−.06
	Indirect						
	$E \rightarrow M \rightarrow F$	—	—	—	—	kp	−.10
	Total	—	—	f	−.16	f	−.16

[a]This table summarizes information given in the path diagrams in Figures 4.1 and 4.2. To minimize rounding errors, the numbers in the table were calculated from values of the coefficients specified to four decimal places [as given in equations (4.10)–(4.15)], rather than two decimal places as shown in Figure 4.2. Because E is endogenous in models (b) and (c), the direct, indirect, and total effects of E all incorporate a control for U. Note that $b = g + df = r + iq + mp + ikp$ and $f = q + kp$.

4.2. PATH MODELS WITH MORE THAN ONE EXOGENOUS VARIABLE

A somewhat more complicated model, with ethnicity ($I = 1$ if Indian, 0 if Fijian) added as a second exogenous variable, is shown in Figure 4.3. In Figure 4.3, *the double-headed curved arrow indicates that U and I may be correlated. No assumption is made about the nature of causation between these two variables.*

The equations for the three estimated models shown in Figure 4.3 are

Model (a)

$$\hat{F} = a + bU + cI \tag{4.16}$$

Model (b)

$$\hat{E} = d + eU + fI \tag{4.17}$$

$$\hat{F} = g + hE + iU + jI \tag{4.18}$$

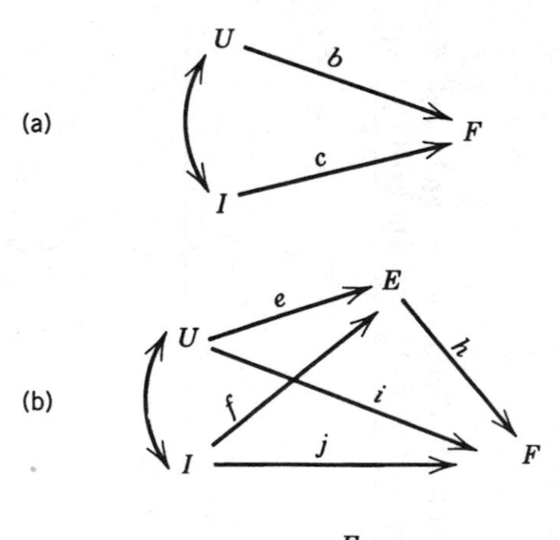

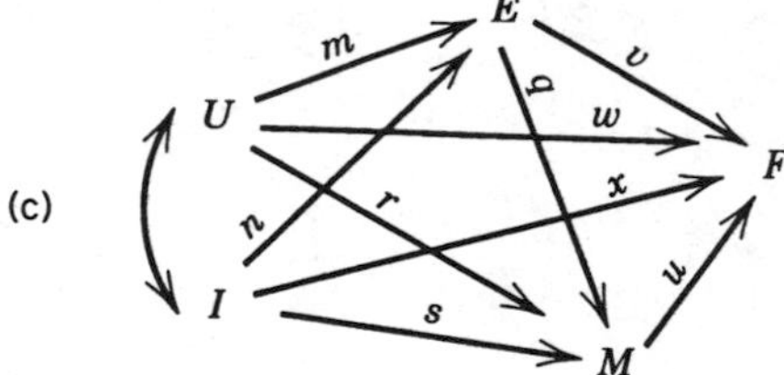

Figure 4.3. Path diagrams for modeling the effects of ethnicity, residence, education, and age at first marriage on fertility. **Note:** The equations for these models are given in equations (4.16)–(4.21) in the text.

Model (c)

$$\hat{E} = k + mU + nI \tag{4.19}$$

$$\hat{M} = p + qE + rU + sI \tag{4.20}$$

$$\hat{F} = t + uM + vE + wU + xI \tag{4.21}$$

In model (b), we obtain the reduced form of the model by substituting (4.17) into (4.18), yielding

$$\hat{F} = (g + dh) + (i + eh)U + (j + fh)I \tag{4.22}$$

The total effect of U on F is the sum of a direct effect, i, and an indirect effect, eh. Because (4.22) and (4.16) have the same mathematical form, the total effect of U on F is the same in models (a) and (b), so that $b = i + eh$. *The direct, indirect, and total effects of U on F all incorporate a control for I.* (Again, because of this control, some authors do not refer to $i + eh$ as a total effect.)

Similarly, the total effect of I on F is the sum of a direct effect, j, and an indirect effect, fh. The total effect of I on F is the same in models (a) and (b), so that $c = j + fh$. The direct, indirect, and total effects of I on F all incorporate a control for U.

Turning now to model (c), we obtain the partially reduced form of the model by substituting (4.20) into (4.21), yielding

$$\hat{F} = (t + pu) + (v + qu)E + (w + ru)U + (x + su)I \tag{4.23}$$

This equation, which eliminates the intervening variable between E and F, says that the total effect of E on F is the sum of a direct effect v (via the path $E \rightarrow F$) and an indirect effect qu (via the path $E \rightarrow M \rightarrow F$), where all three effects incorporate controls for U and I because these latter two variables also appear in the equation. The coefficients of U and I in (4.23) are of less interest than the coefficient of E because the former coefficients omit effects of U and I on F via E. Because equation (4.23) has the same form as equation (4.18), it is clear that the total effect of E on F is the same in models (b) and (c), so that $h = v + qu$.

The fully reduced form of model (c) is obtained by substituting (4.19) into (4.23), yielding

$$\hat{F} = (t + pu + kv + kqu) + (w + mv + ru + mqu)U + (x + nv + su + nqu)I \qquad (4.24)$$

The total effect of U on F is given by the coefficient of U, consisting of a direct effect w (via the path $U \to F$) and indirect effects mv (via the path $U \to E \to F$), ru (via the path $U \to M \to F$), and mqu (via the path $U \to E \to M \to F$). The direct, indirect, and total effects of U on F all incorporate a control for I because I appears as a separate variable in (4.24). Similarly, the total effect of I is given by the coefficient of I in (4.24), consisting of a direct effect x (via the path $I \to F$) and indirect effects nv (via the path $I \to E \to F$), su (via the path $I \to M \to F$), and nqu (via the path $I \to E \to M \to F$). The direct, indirect, and total effects of I on F all incorporate a control for U, because U appears as a separate variable in (4.24).

We can generalize from the above example as follows: When we look at the direct, indirect, and total effects of an intervening endogenous variable on the response variable, all three of these effects incorporate controls for causally prior variables in the model.

When the equations for models (a), (b), and (c) in Figure 4.3 are fitted to the data, the result is

Model (a)

$$\hat{F} = 5.7147 - .9630U + .9591I \qquad (4.25)$$

Model (b)

$$\hat{E} = 5.5493 + 1.1793U - 3.6171I \qquad (4.26)$$

$$\hat{F} = 6.3754 - .1191E - .8226U + .5285I \qquad (4.27)$$

Model (c)

$$\hat{E} = 5.5493 + 1.1793U - 3.6171I \qquad (4.28)$$

$$\hat{M} = 18.0979 + .2814E - .0642U - 2.3623I \qquad (4.29)$$

$$\hat{F} = 10.2216 - .2125M - .0592E - .8363U + .0264I \qquad (4.30)$$

Figure 4.4 shows path diagrams for the fitted models, and Table 4.2 summarizes direct, indirect, and total effects of U on F, I on F, and E on F. The total effects of U on F are the same in all three models; the total effects of I on F are the same in all three models; and the total effects of E on F are the same in models (b) and (c).

If, however, we compare Figure 4.4c and Table 4.2, on the one hand, with Figure 4.2c and Table 4.1, on the other, we see some differences. When ethnicity is added as an exogenous variable, the direct, indirect, and total effects of U on F and E on F change, because they now incorporate a control for I that previously was lacking.

Table 4.2 shows that the effect of U on F, holding I constant, is mostly direct; the effect of I on F, holding U constant, is almost entirely indirect through E and M; and the effect of E on F, holding both U and I constant, is half direct and half indirect through M.

If, in model (c), we were to look at the equation for F by itself [equation (4.30)], we might erroneously conclude that ethnicity has virtually no effect on fertility, once residence is controlled, because the coefficient of I is only .03. But the coefficient of I in this equation

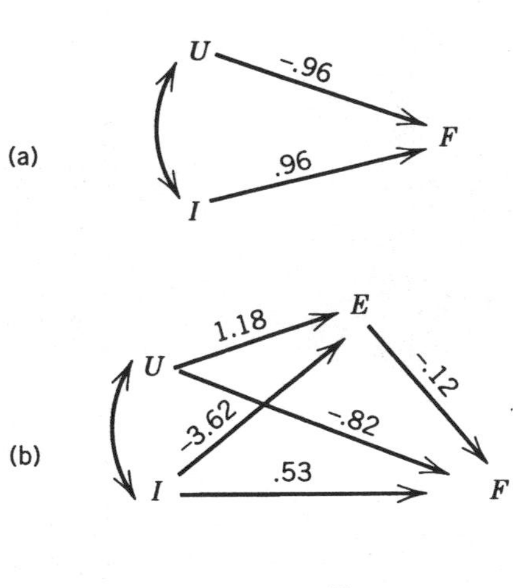

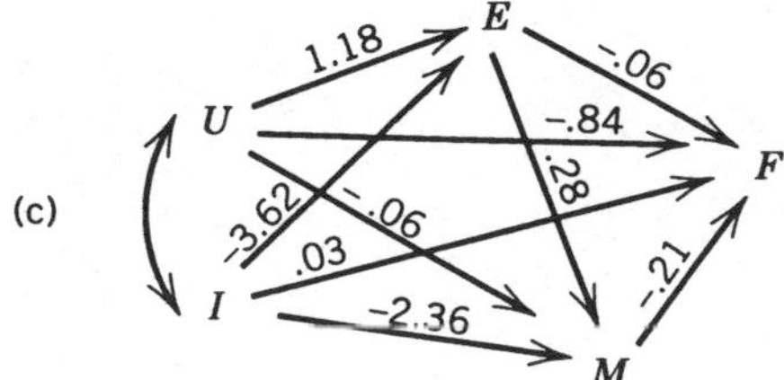

Figure 4.4. Fitted path models of the effects of ethnicity, residence, education, and age at first marriage on fertility: women aged 35–49, 1974 Fiji Fertility Survey. **Note**: The models in this figure are fitted versions of the models in Figure 4.3.

TABLE 4.2. Direct, Indirect, and Total Effects of Residence, Ethnicity, and Education on Fertility: Ever-Married Women Aged 35–49, 1974 Fiji Fertility Survey[a]

		Model					
		(a)		(b)		(c)	
Variable	Type of Effect	Formula	Numerical Value	Formula	Numerical Value	Formula	Numerical Value
Residence	Direct	b	−.96	i	−.82	w	−.84
	Indirect						
	$U \rightarrow E \rightarrow F$	—	—	eh	−.14	mv	−.07
	$U \rightarrow M \rightarrow F$	—	—	—	—	ru	.01
	$U \rightarrow E \rightarrow M \rightarrow F$	—	—	—	—	mqu	−.07
	Total	b	−.96	b	−.96	b	−.96
Ethnicity	Direct	c	.96	j	.53	x	.03
	Indirect						
	$I \rightarrow E \rightarrow F$	—	—	fh	.43	nv	.22
	$I \rightarrow M \rightarrow F$	—	—	—	—	su	.50
	$I \rightarrow E \rightarrow M \rightarrow F$	—	—	—	—	nqu	.21
	Total	c	.96	c	.96	c	.96
Education	Direct	—	—	h	−.12	v	−.06
	Indirect						
	$E \rightarrow M \rightarrow F$	—	—	—	—	qu	−.06
	Total	—	—	h	−.12	h	−.12

[a]This table summarizes information given in the path diagrams in Figures 4.3 and 4.4. To minimize rounding errors, the numbers in the table were calculated from values of the coefficients specified to four decimal places [as given in equations (4.25)–(4.30)], rather than two decimal places as shown in Figure 4.4. The panel for U incorporates a control for I; the panel for I incorporates a control for U; and the panel for E incorporates controls for both U and I. Note that $b = i + eh = w + mv + ru + mqu$, $c = j + fh = x + nv + su + nqu$, and $h = v + qu$.

represents only the direct effect of ethnicity on fertility. The path analytic model shows that the indirect effect of ethnicity on fertility is large, amounting to 0.96 child.

4.3. PATH MODELS WITH CONTROL VARIABLES

A difficulty with the path models considered so far is that some women, particularly those not much above age 35, may still have more children, so that F (number of children ever born to a woman in the age range 35–49) does not necessarily measure completed fertility. Another difficulty is that fertility in Fiji began to decline before 1974 when the survey was taken, so that younger women may have lower completed fertility than older women simply because of the secular trend.

The addition of age as a third exogenous variable, as in Figure 4.5, is one way of dealing with these difficulties. Because the age variable simultaneously captures both a woman's age and her date of birth, it controls for both age and time trend. Insofar as it mixes the two kinds of effects, the meaning of age as a causal variable (i.e., the interpretation of the coefficient of A) is somewhat ambiguous. We shall therefore view age as a control variable rather than as a causal variable, even though it is formally treated like a causal variable in the model.

For simplicity, Figure 4.5 omits the reduced models and shows only the full model. In general form, the equations for the full model are

$$\hat{E} = a + bU + cI + dA \tag{4.31}$$

$$\hat{M} = e + fE + gU + hI + iA \tag{4.32}$$

$$\hat{F} = j + kM + mE + nU + pI + qA \tag{4.33}$$

The fitted versions of these equations are

$$\hat{E} = 10.6383 + 1.2329U - 3.6297I - .1243A \tag{4.34}$$

$$\hat{M} = 17.8271 + .2831E - .0689U - 2.355I + .0064A \tag{4.35}$$

$$\hat{F} = 3.9301 - .2135M - .0193E - .9472U + .1827I + .1487A \tag{4.36}$$

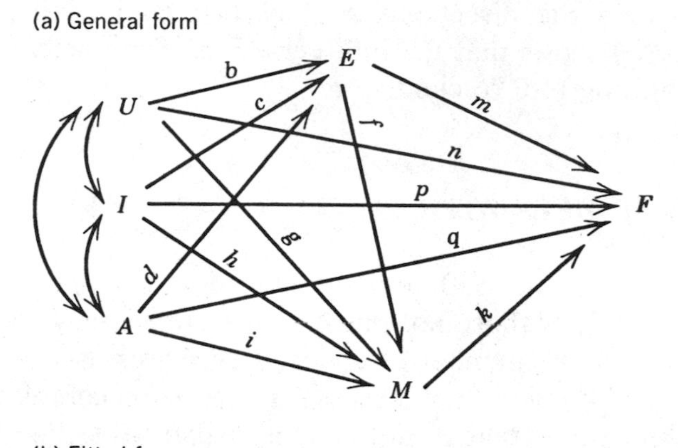

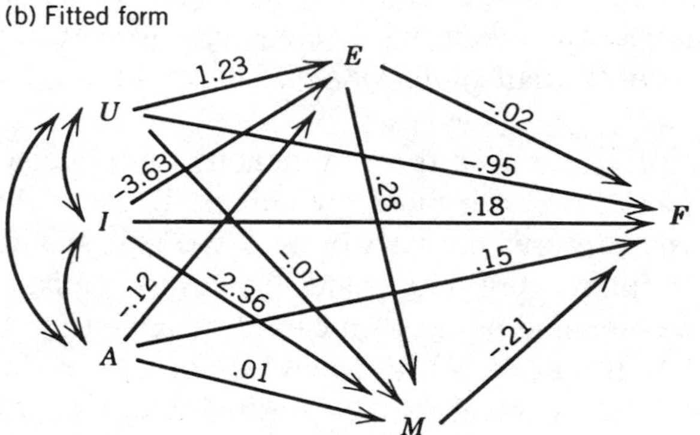

Figure 4.5. Path model of the effects of age, ethnicity, residence, education, and age at first marriage on fertility: women aged 35–49, 1974 Fiji Fertility Survey. **Note:** The equations for this model are given in equations (4.31)–(4.36).

Referring to equations (4.31)–(4.33) and Figure 4.5a, we see that the direct, indirect, and total effects of U on F are n, $bm + gk + bfk$, and $n + bm + gk + bfk$, respectively, where all three effects incorporate controls for I and A. The direct, indirect, and total effects of I on F are p, $cm + hk + cfk$, and $p + cm + hk + cfk$, where all three effects incorporate controls for U and A. The direct, indirect, and total effects of E on F are m, fk, and $m + fk$, respectively, where all three effects incorporate controls for U, I, and A. Because A functions only as a control variable, we do not bother to compute direct, indirect, and total effects for A.

These effects are summarized in Table 4.3. Comparing with Table 4.2, we see that the additional control for A marginally changes the effects of U, I, and E on F.

TABLE 4.3. Direct, Indirect, and Total Effects of Residence, Ethnicity, and Education on Fertility, Incorporating a Control for Age: Ever-Married Women Aged 35–49, 1974 Fiji Fertility Survey[a]

Variable	Type of Effect	Effect	
		Formula	Numerical Value
Residence	Direct	n	−.95
	Indirect		
	$U \to E \to F$	bm	−.02
	$U \to M \to F$	gk	.01
	$U \to E \to M \to F$	bfk	−.07
	Total	$n + bm + gk + bfk$	−1.03
Ethnicity	Direct	p	.18
	Indirect		
	$I \to E \to F$	cm	.07
	$I \to M \to F$	hk	.50
	$I \to E \to M \to F$	cfk	.22
	Total	$p + cm + hk + cfk$	.97
Education	Direct	m	−.02
	Indirect		
	$E \to M \to F$	fk	−.06
	Total	$m + fk$	−.08

[a]This table summarizes information given in the path diagrams in Figure 4.5. To minimize rounding errors, the numbers in this table were calculated from values of the coefficients specified to four decimal places (as given in equations (4.34)–(4.36), rather than two decimal places as shown in Figure 4.5. Age is controlled throughout the table. The panel for U incorporates controls for I and A; the panel for I incorporates controls for U and A; and the panel for E incorporates controls for U, I, and A.

4.4. SATURATED AND UNSATURATED PATH MODELS

In the path models we have considered so far, each endogenous variable depends on all causally prior variables. Pictorially, this means that there is an arrow from each causally prior variable to the endogenous variable in question. Such models are called *saturated models*. Models in which some of the arrows are missing (the coefficients corresponding to the missing arrows are presumed to be zero) are called *unsaturated models*.

Normally one does not use an unsaturated model unless our theory specifies that certain coefficients are zero. The theory can be tested by first running the saturated model and testing whether the coefficients in question differ significantly from zero. If they do not, certain prior variables can be omitted from some or all of the model equations, and the model can be reestimated with some arrows omitted.

The calculation of direct and indirect effects proceeds as before, by taking products of coefficients over each remaining path or by computing the coefficient of the appropriate reduced form of the model. The reduced form may be somewhat different than it was in the saturated model, insofar as one or more variables may be omitted from some of the equations.

For purposes of illustration, suppose that, in the model in Figure 4.5, our theory predicts that the effect of ethnicity on fertility is entirely indirect. In fact it turns out that the direct effect of I on F, 0.18, is not statistically significant at the 5 percent level. If we omit the direct path between I and F from the model, so that the arrow from I to F is dropped from the path diagram, the equations for the resulting unsaturated model are

$$\hat{E} = a + bU + cI + dA \tag{4.37}$$

$$\hat{M} = e + fE + gU + hI + iA \tag{4.38}$$

$$\hat{F} = j + kM + mE + nU + qA \tag{4.39}$$

These equations are the same as (4.31)–(4.33), except that the term pI is omitted in (4.39).

The fitted versions of (4.37)–(4.39) are

$$\hat{E} = 10.6383 + 1.2329U - 3.6297I - .1243A \tag{4.40}$$

$$\hat{M} = 17.8271 + .2831E - .0689U - 2.3555I + .0064A \tag{4.41}$$

$$\hat{F} = 4.2279 - .2180M - .0322E - .9121U + .1468A \tag{4.42}$$

Equations (4.40) and (4.41) are identical to equations (4.34) and (4.35), respectively, but (4.42) differs from (4.36) because of omission of the ethnicity term, which means that the equation had to be reestimated. The coefficients of the remaining predictor variables in (4.42) are nevertheless about the same as they were in (4.36), as expected. The path diagram and table of effects, based on equations (4.40)–(4.42), accordingly differ little from Figure 4.5b and Table 4.3 (aside from the omission of the direct path between I and F) and are not shown.

In the remainder of this chapter we shall consider only saturated models.

4.5. PATH ANALYSIS WITH STANDARDIZED VARIABLES

4.5.1. Standardized Variables and Standardized Path Coefficients

The original formulation of path analysis uses standardized variables, rather than variables in their original metric. Suppose that we wish to reformulate the model

$$\hat{Y} = a + bX \tag{4.43}$$

in standardized form. We first recall that the point $(\overline{X}, \overline{Y})$ falls on the regression line, so that

$$\overline{Y} = a + b\overline{X} \tag{4.44}$$

Subtracting (4.44) from (4.43), we obtain

$$\hat{Y} - \overline{Y} = b(X - \overline{X}) \tag{4.45}$$

If we now divide both sides of (4.45) by s_Y, the standard deviation of Y, and if we both multiply and divide the right side of (4.45) by s_X, we obtain

$$\frac{\hat{Y} - \overline{Y}}{s_Y} = b\frac{s_X}{s_Y}\frac{X - \overline{X}}{s_X} \tag{4.46}$$

We call $(Y - \overline{Y})/s_X$ and $(X - \overline{X})/s_X$ *standardized variables*, and we call $b(s_X/s_Y)$ a *standardized regression coefficient*.

Denoting standardized variables by lowercase italicized Roman letters and the standardized coefficient as p_{yx}, we can rewrite (4.46) in *standardized form* as

$$\hat{y} = p_{yx}x \tag{4.47}$$

where

$$\boxed{p_{yx} \equiv b\frac{s_x}{s_y}} \tag{4.48}$$

In (4.47) the constant term is zero, and $\hat{y} = 0$ when $x = 0$. Thus the regression line always passes through the origin $(0, 0)$ when the equation is put into standardized form. A change in X, from X to $X + s_X$, translates into a change in x of $\{[(X + s_X) - \overline{X}]/s_X\} - [(X - \overline{X})/s_X] = s_x/s_x = 1$. *Thus a one-unit increase in x corresponds to an increase of one standard deviation in X.*

In original X units, s_X is a measure of the spread of the distribution of X in the sample. Thus values of $x = (X - \overline{X})/s_X$ are *distribution-dependent*, whereas the original values of X are not. Similarly, the interpretation of $p_{yx} = b(s_x/s_Y)$ depends on the distributions of both X and Y. The coefficient p_{yx} is the effect on y of a one-unit increase in x; equivalently, it is the effect on Y (measured in standard deviation units) of an increase of one standard deviation in X.

For a model with two predictor variables we have similarly that

$$\hat{Z} = a + bX + cY \tag{4.49}$$

$$\overline{Z} = a + b\overline{X} + c\overline{Y} \tag{4.50}$$

$$\hat{Z} - \overline{Z} = b(X - \overline{X}) + c(Y - \overline{Y}) \tag{4.51}$$

$$\frac{\hat{Z} - \overline{Z}}{s_Z} = b\frac{s_X}{s_Z}\frac{X - \overline{X}}{s_X} + c\frac{s_Y}{s_Z}\frac{Y - \overline{Y}}{s_Y} \tag{4.52}$$

which can be written as

$$\hat{z} = p_{zx}x + p_{zy}y \tag{4.53}$$

In the bivariate case in (4.46), *the standardized coefficient* $b(s_X/s_Y)$ *is the same as the simple bivariate coefficient of correlation between X and Y*, as discussed earlier in Chapter 1 [see equation (1.36)]. In the multivariate case with two predictor variables in (4.52), however, the standardized coefficients do not have such a simple interpretation.

4.5.2. Path Models in Standardized Form

Conventional path analysis is associated with a particular notation that is illustrated in Figure 4.6. Figure 4.6 is similar to Figures 4.1c and Figure 4.2c, except for notation, standardization of variables, and the inclusion of the *residual variables* x_u, x_v, and x_w. The equations

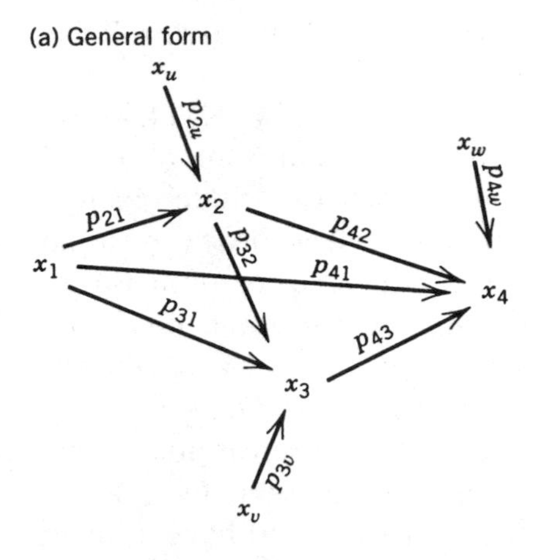

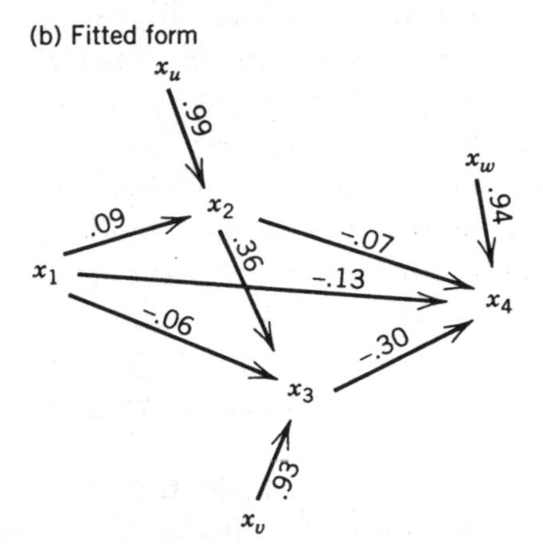

Figure 4.6. An example of path analysis using standardized variables: women aged 35–39, 1974 Fiji Fertility Survey. **Note:** The models in this figure are standardized versions of the models in Figures 4.1c and 4.2c. x_1 is residence, x_2 is education, x_3 is age at first marriage, and x_4 is fertility.

for the model in Figure 4.6 are

$$x_2 = p_{21}x_1 + p_{2u}x_u \tag{4.54}$$

$$x_3 = p_{31}x_1 + p_{32}x_2 + p_{3v}x_v \tag{4.55}$$

$$x_4 = p_{41}x_1 + p_{42}x_2 + p_{43}x_3 + p_{4w}x_w \tag{4.56}$$

The standardized variables are represented by lowercase italicized Roman letters. The response variable is denoted by x_4 and the predictor variables by x_1, x_2, and x_3, where the ordering x_1, x_2, x_3, x_4 represents a causal ordering.

The path coefficients (i.e., the standardized regression coefficients) are denoted by p_{ji}, where, in terms of an arrow in the path diagram, x_i

is at the base of the arrow and x_j is at the head of the arrow. Thus one can read ji as "to j from i."

Note that the response variables x_2, x_3, and x_4 in (4.54)–(4.56) do not have hats over them. This means that they represent observed rather than estimated values, so that it is necessary to include residual terms on the right sides of these equations. These extra residual terms, $p_{2u}x_u$, $p_{3v}x_v$, and $p_{4w}x_w$, are conceptualized as unmeasured sources of variation in the response variable in the same equation. Each residual term is conceptualized as the product of a coefficient and a standardized composite variable, called a *residual variable*, representing unmeasured determinants of the response variable in the same equation. The residual variable, which is denoted by a letter instead of a number in the subscript of x, is assumed to have a mean of zero and a standard deviation of one, and to be uncorrelated with other predictor variables in the model. *This means that the residual term behaves mathematically like an ordinary residual, e. Therefore, we can estimate equations* (4.54)–(4.56) *by ordinary least squares*. In practice, each standardized coefficient is obtained by multiplying the corresponding unstandardized coefficient by the ratio of the standard deviation of the corresponding predictor variable to the standard deviation of the response variable, as shown earlier in equations (4.49), (4.52), and (4.53). In BMDP, SAS, and SPSS, programs for regression contain options for printing out standardized as well as unstandardized regression coefficients.

However, multiple regression programs do not print out the coefficient of the residual variable in each equation. It can be shown (Duncan, 1975, Chapter 4) that

$$p_{2u} = \sqrt{1 - r_{21}^2} = \sqrt{1 - R_{21}^2} \tag{4.57}$$

$$p_{3v} = \sqrt{1 - R_{321}^2} \tag{4.58}$$

$$p_{4w} = \sqrt{1 - R_{4321}^2} \tag{4.59}$$

where r_{21} (equal in absolute magnitude to R_{21}) denotes the bivariate correlation between x_1 and x_2, R_{321} denotes the multiple correlation between x_3, x_2, and x_1, and R_{4321} denotes the multiple correlation between x_4, x_3, x_2, and x_1. The quantity $1 - R^2$ is called the *residual variance*. Because R^2 is printed out as part of the OLS regression results for each model equation, it is a simple matter to calculate p_{2u}, p_{3v}, and p_{4w} using (4.57)–(4.59).

In general, the path coefficients of the residual variables are calculated as $\sqrt{1 - R^2}$, where R^2, which is not adjusted, pertains to the equation corresponding to the endogenous variable to which the residual path points. We see that the residual path coefficient is a simple transformation of R^2 for the equation in which the residual variable appears. *Thus the residual path coefficient may be viewed as a way of incorporating* R^2 *into the path diagram*. Because $1 - R^2 =$ USS/TSS, $\sqrt{1 - R^2}$ is simply the square root of the proportion of the total sum of squares that is not explained by the equation (i.e., the square root of the residual variance). The residual path coefficient therefore ranges between 0 and 1. *If the residual path coefficient is close to* 1, *the equation explains very little of the variation in the response variable. If it is close to* 0, *the equation accounts for almost all of the variation in the response variable.* The residual path coefficients in Figure 4.6b indicate that the equations of the model do not explain much of the variation in the endogenous variables.

We cannot incorporate residual path variables and residual path coefficients in the comparable unstandardized path diagrams in Figures 4.1c and 4.2c, because we have no way of knowing the standard deviations of the residual variables.

4.5.3. Standardized Versus Unstandardized Coefficients

Both standardized and unstandardized coefficients are useful, but most of the time we prefer the unstandardized coefficients because their meaning is clearer. Sometimes we have variables, such as a socioeconomic index variable, where the variable is formed as a sum of item scores, so that the metric does not have a clear meaning. In this case it may be useful to standardize the variable so that the metric is defined in terms of the spread of the distribution as measured in standard deviation units. But in the case of variables like rural–urban residence or completed years of education, the original metric has a clear meaning that should not be muddied by standardizing the variable, thereby making it distribution-dependent. *There is no problem with including both standardized and unstandardized variables in the same path model.* Of course, when some variables are not standardized, we cannot compute residual path coefficients for equations that include unstandardized variables.

Standardized coefficients are also useful when we wish to compare the effects of predictor variables that are measured in different units. If the predictor variables are education and income, for example, education might be measured as years of completed education, and income might be measured in dollars. The unstandardized coefficients of these variables are noncomparable, because the coefficients measure change in the response variable per unit increase in the predictor variable and because the units of these two predictor variables are noncomparable. The standardized coefficients, on the other hand, are comparable because they are both measured in standard deviation units that do not depend on the original units of measurement. Note, however, that standardized coefficients are not comparable across two or more samples, because the distributions and therefore the standard deviations of the variables usually differ among the samples.

4.6. PATH MODELS WITH INTERACTIONS AND NONLINEARITIES

It is possible to incorporate interactions and nonlinearities into path models. To illustrate how this is done, we return to the unstandardized form.

As an example, let us consider in more detail the model shown earlier in Figure 4.5. Suppose that we have reason to believe that ethnicity interacts with other predictor variables in the model. The easiest way to test for this interaction is to run the model separately for Indians and Fijians. If the coefficients of the predictor variables are about the same in the separate models, then there is no interaction. But if some of the coefficients are quite different, then interaction is present and we must modify the model.

The separate models for Indians and Fijians are shown in Figure 4.7. By this time, the reader should be able to reconstruct the underlying equations for each of these two models, so these equations are not shown. Figure 4.7 shows that the direct effect of E on F does not differ greatly between Indians and Fijians. The same is true of the direct effect of M on F, the direct effect of E on M, the direct effect of A on F, the direct effect of A on M, and the direct effect of A on E. On the other hand, the effects of U on E, M, and F differ substantially between Indians and Fijians, indicating the presence of a UI interaction that should be taken into account in the model.

To take into account this interaction, we modify the model in Figure 4.5 by adding UI terms in the model equations. For purposes of illustration, we shall also add a nonlinearity in the effect that age at first marriage has on fertility. The revised model equations in general

(a) Indians

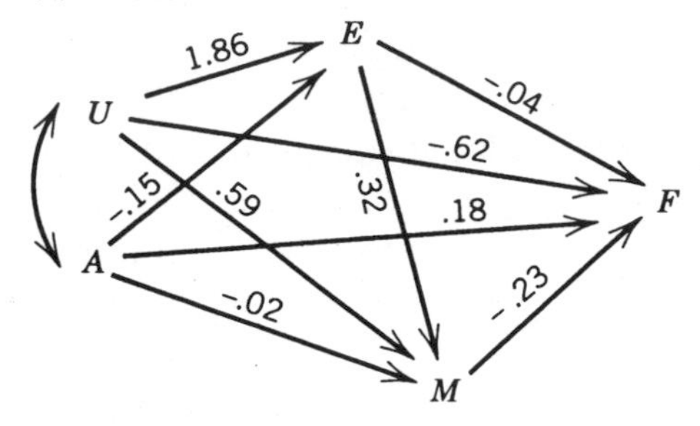

(b) Fijians

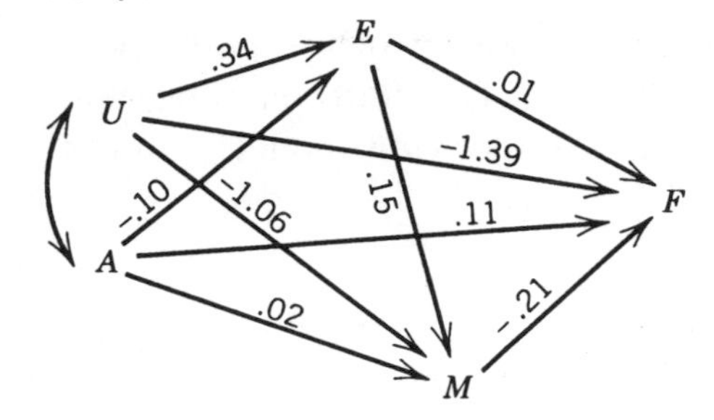

Figure 4.7. Path models of the effects of age, residence, education, and age at first marriage on fertility, by ethnicity: women aged 35–49, 1974 Fiji Fertility Survey. **Note**: Variables and path coefficients are not standardized.

form are

$$\hat{E} = a + bU + cI + dUI + eA \tag{4.60}$$

$$\hat{M} = f + gE + hU + iI + jUI + kA \tag{4.61}$$

$$\hat{F} = m + nM + pM^2 + qE + rU + sI + tUI + vA \tag{4.62}$$

In fitted form the equations are

$$\hat{E} = 11.0169 + .3435U - 4.1214I + 1.5000UI - .1276A \tag{4.63}$$

$$\hat{M} = 18.4770 + .2641E - 1.0970U - 3.0059I + 1.7737UI + .0001A \tag{4.64}$$

$$\hat{F} = 3.7312 - .1615M - .0014M^2 - .0277E - 1.3884U - .0920I + .7559UI + .1463A \tag{4.65}$$

Recall from Chapter 2 that, in the presence of interaction between residence and ethnicity, the effect of U on E is $b + dI$, and the effect of I on E is $c + dU$. The effects of U and I on other variables in the model have a similar compound form. Recall also from Chapter 2 that, in the presence of a nonlinear effect of age at first marriage modeled by a quadratic term, the effect of M on F is $n + 2pM$. The path coefficients are shown in general form in Figure 4.8a and in fitted form in Figure 4.8b.

The results may also be presented in tabular form, as in Table 4.4. Table 4.4 has the same form as Table 4.3 and is formed in the same way. Direct effects are simply the path coefficients for the direct paths, and the indirect effects are products of path coefficients along the indirect paths.

If a value of M is specified, say $M = \overline{M}$, one can redraw Figure 4.8b as four separate diagrams, one each for rural Fijians, rural Indians, urban Fijians, and urban Indians, by setting $(U, I) = (0, 0)$, $(0, 1)$, $(1, 0)$, and $(1, 1)$ in the formulae for certain of the path coefficients in the path diagram in Figure 4.8b. The four separate path diagrams are all derived from the same underlying model. Alterna-

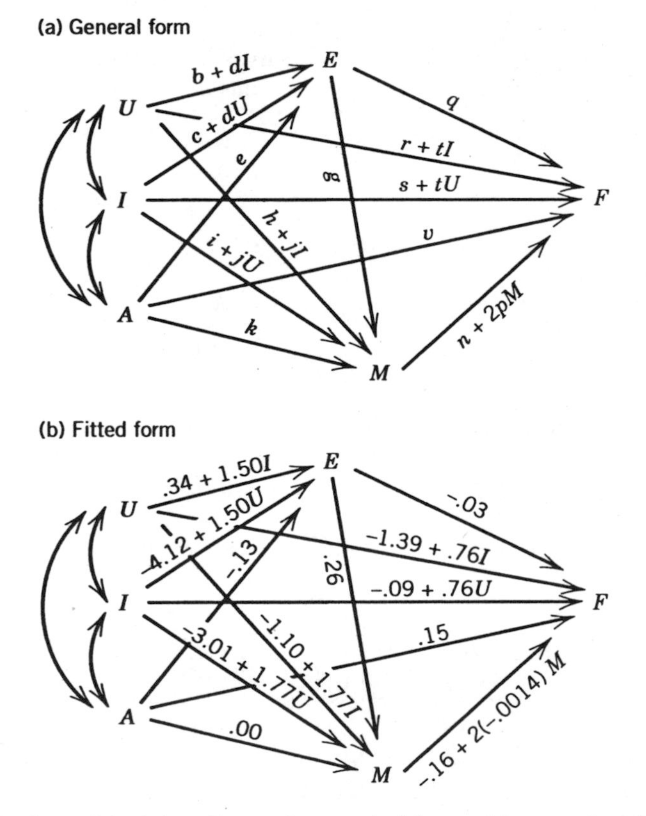

Figure 4.8. Path model of the effects of age, ethnicity, residence, education, and age at first marriage on fertility: women aged 35–49, 1974 Fiji Fertility Survey. **Note**: The equations for this model are given in equations (4.63)–(4.65).

TABLE 4.4. Direct, Indirect, and Total Effects of Residence, Ethnicity, and Education on Fertility, with Age Controlled and Incorporating Interaction between Residence and Ethnicity and Nonlinear Effects of Age at First Marriage: Ever-Married Women Aged 35–49, 1974 Fiji Fertility Survey[a]

Variable	Type of Effect	Effect: Formula	Effect: Numerical Value
Residence	Direct	$r + tI$	$-1.39 + .76I$
	Indirect		
	$U \to E \to F$	$(b + dI)q$	$(.34 + 1.50I)(-.03)$
	$U \to M \to F$	$(h + jI)(n + 2pM)$	$(-1.10 + 1.77I)(-.16 - .003M)$
	$U \to E \to M \to F$	$(b + dI)g(n + 2pM)$	$(.34 + 1.50I)(.26)(-.16 - .003M)$
	Total	Sum of direct and indirect	Sum of direct and indirect
Ethnicity	Direct	$s + tU$	$-.09 + .76U$
	Indirect		
	$I \to E \to F$	$(c + dU)q$	$(-4.12 + 1.50U)(-.03)$
	$I \to M \to F$	$(i + jU)(n + 2pM)$	$(-3.01 + 1.77U)(-.16 - .003M)$
	$I \to E \to M \to F$	$(c + dU)g(n + 2pM)$	$(-4.12 + 1.50U)(.26)(-.16 - .003M)$
	Total	Sum of direct and indirect	Sum of direct and indirect
Education	Direct	q	$-.03$
	Indirect		
	$E \to M \to F$	$g(n + 2pM)$	$.26(-.16 - .003M)$
	Total	$q + g(n + 2pM)$	$-.03 + .26(-.16 - .003M)$

[a] This table summarizes information given in the path diagrams in Figure 4.8. Age is controlled throughout the table.

tively, one could also set $U = \bar{U}$ and then construct two diagrams, one for Fijians and one for Indians by setting I alternatively to 0 and 1. Again, the two diagrams are derived from the same underlying interactive model, unlike the two diagrams in Figure 4.7, which are derived from two separate noninteractive models.

In the same way, Table 4.4 can be disaggregated into separate tables for rural Fijians, rural Indians, urban Fijians, and urban Indians, or, alternatively, into separate tables for Fijians and Indians. Thus, despite the more complicated nature of the model, we can present it in a series of path diagrams and tables that are just as easy to read as before.

Why bother to do this? Why not use separate models? When we use separate models, we greatly reduce the number of cases on which each model is based. Coefficients are then less likely to be statistically significant. If the combined model applied to the entire sample fits well, it has more statistical power.

4.7. FURTHER READING

For concise, readable introductions to path analysis, see Kendall and O'Muircheartaigh (1977) and Asher (1983). For a more comprehensive treatment, see Duncan (1975). See Holland (1988) for an instructive discussion of causal inference in path analysis.

5

LOGIT REGRESSION

Logistic regression, more commonly called *logit regression*, is used when the response variable is dichotomous (i.e., binary or 0–1). The predictor variables may be quantitative, categorical, or a mixture of the two.

Sometimes ordinary bivariate or multiple regression is used in this situation. When this is done, the model is called the *linear probability model*. The linear probability model provides a useful introduction to the logit regression model, so we consider it first.

5.1. THE LINEAR PROBABILITY MODEL

Suppose that we wish to analyze the effect of education on current contraceptive use among fecund, nonpregnant, currently married women aged 35–44. Our variables are

C: contraceptive use (1 if using, 0 otherwise)
E: number of completed years of education

The scatterplot of C against E, for a few representative points, might look something like the plot in Figure. 5.1. An ordinary bivariate regression line can be fitted through the points by ordinary least squares (OLS), as shown. The estimated equation of the line has the

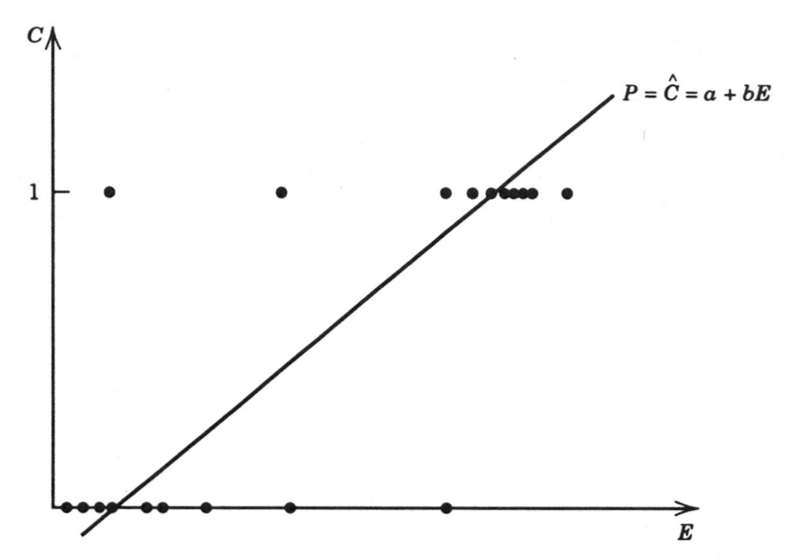

Figure 5.1. The linear probability model: probability of contraceptive use predicted by education. **Notes:** C denotes contraceptive use (1 if using, 0 otherwise), and E denotes number of completed years of education. $\hat{C}$, denoting the value of C predicted from the regression, is interpreted as the probability of currently using contraception.

form

$$\hat{C} = a + bE \tag{5.1}$$

where $\hat{C}$ is the value of C predicted by the regression.

The observed value of C can assume only two values, 0 and 1. In contrast, the value of C predicted by the regression line, $\hat{C}$, can assume a continuum of values. For most observed values of the predictor variable, $\hat{C}$ will have a value between 0 and 1. *We interpret $\hat{C}$ as the probability that a woman with a specified level of education is currently using contraception.*

Because $\hat{C}$ is interpreted as a probability, (5.1) may also be written as

$$P = a + bE \tag{5.2}$$

where P denotes the estimated probability of use. If so desired, (5.2) can be elaborated by adding more predictor variables, interaction terms, quadratic terms, and so on.

Although the linear probability model has the advantage of simplicity, it suffers from some serious disadvantages:

1. *The estimated probability P can assume impossible values.* At the lower left end of the line in Figure 5.1, P is negative, and at the upper right end, P exceeds unity.
2. *The linearity assumption is seriously violated.* According to this assumption, the expected value of C at any given value of E falls on the regression line. But this is not possible for the parts of the line for which $P < 0$ or $P > 1$. In these regions, the observed points are either all above the line or all below the line.
3. *The homoscedasticity assumption is seriously violated.* The variances of the C values tend not to behave properly either. The variance of C tends to be much higher in the middle range of E than at the two extremes, where the values of C are either mostly zeros or mostly ones. In this situation, the equal-variance assumption is untenable.
4. *Because the linearity and homoscedasticity assumptions are seriously violated, the usual procedures for hypothesis testing are invalid.*
5. *R^2 tends to be very low.* The fit of the line tends to be very poor. Because the response variable can assume only two values, 0 and 1, the C values tend not to cluster closely about the regression line.

For these reasons, the linear probability model is seldom used, especially now that alternative models such as logit regression are widely available in statistical software packages such as SAS, LIMDEP, BMDP, and SPSS.

5.2. THE LOGIT REGRESSION MODEL

Instead of a straight line, it seems preferable to fit some kind of sigmoid curve to the observed points. *By a sigmoid curve, we mean a curve that resembles an elongated S or inverted S laid on its side.* The tails of the sigmoid curve level off before reaching $P = 0$ or $P = 1$, so that the problem of impossible values of P is avoided.

Two hypothetical examples of sigmoid curves are shown in Figure 5.2. In Figure 5.2a, the response variable is C, as before, and the curve increases from near zero at low levels of education to near unity

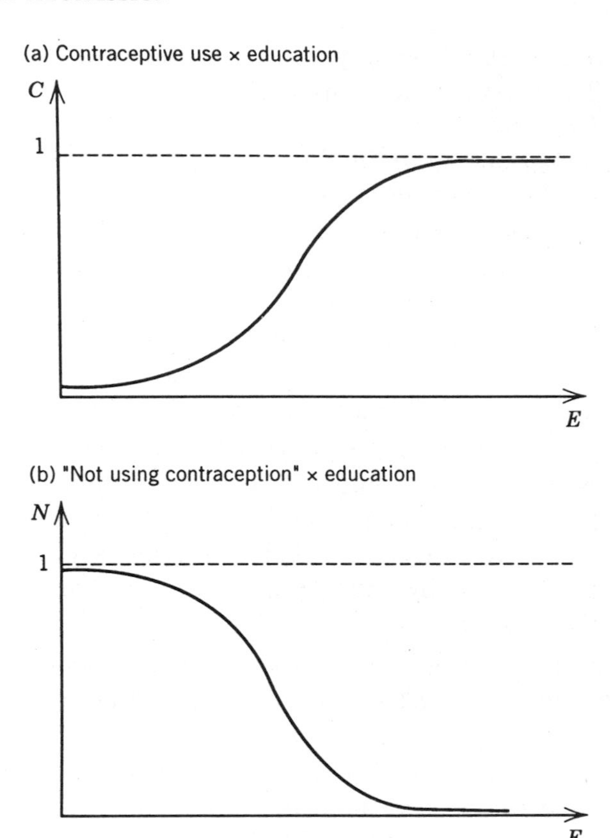

Figure 5.2. Two examples of sigmoid curves.

at high levels of education. In Figure 5.2b, the response variable is "not using contraception," denoted by N, where N is one if the woman is not using contraception and zero otherwise. Now the curve decreases from near unity at low levels of education to near zero at high levels of education.

The sigmoid curve assumes that the predictor variable has its largest effect on P when P equals 0.5, and that the effect becomes smaller in absolute magnitude as P approaches 0 or 1. (The effect of E on P is measured by the slope of a line tangent to the curve at each specified value of E.) In many situations, this is a reasonably accurate portrayal of reality.

5.2.1. The Logistic Function

What mathematical form should we assign to the sigmoid curve? Although there are many possibilities, the logistic function tends to be preferred, partly because it is comparatively easy to work with mathematically, and partly because it leads to a model (the logit regression model) that is comparatively easy to interpret.

There is, however, a certain amount of arbitrariness in the choice of functional form to represent the sigmoid curve, and results from the model depend to some extent on which functional form is chosen. Another functional form that is frequently used is the cumulative normal distribution, which forms the basis of the probit regression model. The logit and probit models usually yield similar but not identical results. We do not consider the probit model any further in this book.

The basic form of the logistic function is

$$\boxed{P = \frac{1}{1 + e^{-Z}}} \tag{5.3}$$

where Z is the predictor variable and e is the base of the natural logarithm, equal to 2.71828 . . . . Throughout this chapter we shall view (5.3) as an estimated model, so that P is an estimated probability.

If numerator and denominator of the right side of (5.3) are multiplied by e^Z, the logistic function in (5.3) can be written alternatively as

$$P = \frac{e^Z}{1 + e^Z} = \frac{\exp(Z)}{1 + \exp(Z)} \tag{5.4}$$

where $\exp(Z)$ is another way of writing e^Z. Equation (5.3), or equivalently (5.4), is graphed in Figure 5.3a.

A property of the logistic function, as specified by (5.3), is that when Z becomes infinitely negative, e^{-Z} becomes infinitely large, so that P approaches 0. When Z becomes infinitely positive, e^{-Z} becomes infinitesimally small, so that P approaches unity. When $Z = 0$, $e^{-Z} = 1$, so that $P = .5$. Thus the logistic curve in Figure 5.3a has its "center" at $(Z, P) = (0, .5)$.

To the left of the point (0, .5), the slope of the curve (i.e., the slope of a line tangent to the curve) increases as Z increases. To the right of this point, the slope of the curve decreases as Z increases. *A point with this property is called an inflection point.*

Suppose that we have a theory that Z is a cause of P. Then the slope of the curve at a particular value of Z measures the effect of Z on P at that particular value of Z. *Therefore, to the left of the*

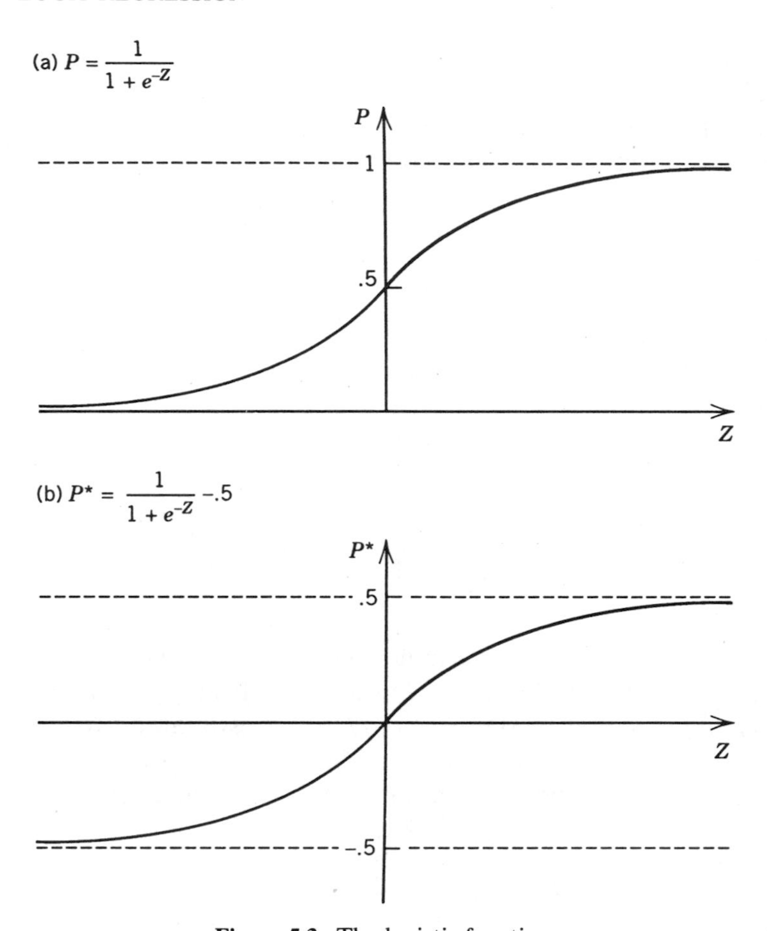

Figure 5.3. The logistic function.

inflection point, the effect of Z on P increases as Z increases. To the right of the inflection point, the effect of Z on P decreases as Z increases. The effect of Z on P attains its maximum at the inflection point. Effects are not constant over the range of the predictor variable, as they are in the simple bivariate regression model.

Another property of the logistic curve is that it is symmetric about its inflection point, as can be demonstrated as follows: First, subtract 0.5 from P, so that the curve is moved downward by 0.5, as shown in Figure 5.3b. The inflection point is now at the origin instead of (0, .5). Except for this downward translation, the shape of the curve is the same as before. Of course, when we do this translation, P can no longer be interpreted as a probability, because P now ranges between -0.5 and $+0.5$. We therefore rename P as P^*. The equation of the curve is now

$$P^*(Z) = \frac{1}{1 + e^{-Z}} - .5 \tag{5.5}$$

where we write P^* as $P^*(Z)$ to emphasize the functional dependence of P^* on Z.

The symmetry of the curve, as graphed in Figure 5.3b, can now be demonstrated by showing that the vertical upward distance from the horizontal axis to the curve at any given positive value of Z equals the vertical downward distance from the horizontal axis to the curve at $-Z$—in other words, by proving that $P^*(Z) = -P^*(-Z)$. This is done by showing that the equation

$$\frac{1}{1 + e^{-Z}} - .5 = -\left[\frac{1}{1 + e^{Z}} - .5\right] \tag{5.6}$$

reduces to an identity, an exercise that is left up to the reader. [An identity is an equation that is true regardless of the value of the unknown—in this case Z. The validity of (5.6) can be demonstrated by reducing (5.6) to the form $0 = 0$.] A less rigorous demonstration that the formula is an identity can be accomplished by substituting in a few randomly chosen values of Z and checking to see that equality between the left and right sides of (5.6) is preserved.

5.2.2. The Multivariate Logistic Function

Equation (5.3) is bivariate. How can we make it multivariate? Suppose that Z, instead of being a single predictor variable, is a linear function of a set of predictor variables:

$$Z \equiv b_0 + b_1X_1 + b_2X_2 + \cdots + b_kX_k \tag{5.7}$$

(Note that Z is not a response variable in this equation.) This expression can be substituted for Z in the formula for the logistic function in (5.3):

$$\boxed{P = \frac{1}{1 + e^{-(b_0 + b_1X_1 + b_2X_2 + \cdots + b_kX_k)}}} \tag{5.8}$$

All the basic properties of the logistic function are preserved when this substitution is done. The function still ranges between 0 and 1 and achieves its maximum rate of change, with respect to change in any of the X_i, at $P = .5$.

As a simple example, suppose that (5.7) assumes the very simple form $Z = -X$. Then (5.8) becomes

$$P = \frac{1}{1 + e^X} \tag{5.9}$$

When $Z = -X$, as in equation (5.9), *the graph of P against X is a reversed sigmoid curve, which is* 1 *at* $-\infty$ *and* 0 *at* $+\infty$.

As a slightly more complicated example, suppose that (5.7) takes the form $Z = a + bX$, where a and b are parameters that are fitted to the data:

$$P = \frac{1}{1 + e^{-(a+bX)}} \tag{5.10}$$

Equation (5.10) can be rewritten as

$$P = \frac{1}{1 + e^{-b(a/b+X)}} \tag{5.11}$$

from which it is evident that the curve is centered at $X = -a/b$ instead of $X = 0$. *The constant term a/b shifts the curve to the left or right, depending on whether a/b is positive or negative, and the coefficient b stretches or compresses the curve along the horizontal dimension, depending on whether $|b|$ (the absolute value of b) is less than or greater than 1. If b is negative, the curve goes from 1 to 0 instead of 0 to 1 as X increases*. (The reader may demonstrate these properties by picking some alternative sets of values of a and b and then constructing, for each set, a graph of P against X for (5.3) with $a + bX$ substituted for Z.) The introduction of the parameters a and b makes the model more flexible, so that a better fit can be achieved.

5.2.3. The Odds and the Logit of *P*

The logit of P is derived from the logistic function

$$P = \frac{1}{1 + e^{-Z}} \tag{5.3 repeated}$$

From (5.3) it follows that

$$1 - P = 1 - \frac{1}{1 + e^{-Z}} = \frac{e^{-Z}}{1 + e^{-Z}} \tag{5.12}$$

Dividing (5.3) by (5.12) yields

$$\frac{P}{1 - P} = e^{Z} \tag{5.13}$$

Taking the natural logarithm (base e) of both sides of (5.13), we obtain

$$\boxed{\log \frac{P}{1 - P} = Z} \tag{5.14}$$

The quantity $P/(1 - P)$ is called the *odds*, denoted more concisely as Ω (uppercase omega), and the quantity $\log[P/(1 - P)]$ is called the *log odds* or the *logit of P*. Thus

$$\boxed{\text{Odds} \equiv \frac{P}{1 - P} \equiv \Omega} \tag{5.15}$$

and

$$\boxed{\text{logit } P \equiv \log \frac{P}{1 - P} \equiv \log \Omega} \tag{5.16}$$

The definition of the odds in (5.15) corresponds to everyday usage. For example, one speaks of the odds of winning a gamble on a horse race as, say, "75 : 25", meaning $.75/(1 - .75)$ or, equivalently, $75/(100 - 75)$. Alternatively, one speaks of "three-to-one" odds, which is the same as 75 : 25.

With these definitions, and with the expression in (5.7) substituted for Z, (5.14) can be written alternatively as

$$\boxed{\text{logit } P = b_0 + b_1 X_1 + b_2 X_2 + \cdots + b_k X_k} \tag{5.17}$$

$$\boxed{\log \frac{P}{1 - P} = b_0 + b_1 X_1 + b_2 X_2 + \cdots + b_k X_k} \tag{5.18}$$

or

$$\boxed{\log \Omega = b_0 + b_1X_1 + b_2X_2 + \cdots + b_kX_k} \tag{5.19}$$

Equations (5.17)–(5.19) are in the familiar form of an ordinary multiple regression equation. This is advantageous, because some of the statistical tools previously developed for multiple regression can now be applied to logit regression.

5.2.4. Logit Regression Coefficients as Measures of Effect on Logit P

The discussion of effects is facilitated by considering a simple specification of (5.17) that can be fitted to data from the 1974 Fiji Fertility Survey. Suppose that we wish to investigate the effect of education, residence, and ethnicity on contraceptive use among fecund, nonpregnant, currently married women aged 35–44. Sample size is $n = 954$. Our variables are

P: estimated probability of contraceptive use
E: number of completed years of education
U: 1 if urban, 0 otherwise
I: 1 if Indian, 0 otherwise

For U, "otherwise" means rural, and for I, "otherwise" means Fijian, as in Chapter 3.

Our model is

$$\text{logit } P = a + bE + cU + dI \tag{5.20}$$

which, when fitted to the Fiji data, is

$$\begin{aligned}\text{logit } P = -\underset{(.181)}{.611} + \underset{(.025)}{.055E} + \underset{(.153)}{.378U} + \underset{(.166)}{1.161I}\end{aligned} \tag{5.21}$$

Numbers in parentheses under the coefficients are standard errors. In every case, the coefficient is at least twice its standard error, indicating that the coefficient differs significantly from zero at the 5 percent level with a two-tailed test. We shall consider later how the fitting is done.

Just as in ordinary multiple regression, the coefficients of the predictor variables can be interpreted as effects if there is theoretical justification for doing so. Because (5.21) is in the form of a multiple regression equation, we can immediately say, for example, that the effect of a one-year increase in education, controlling for residence and ethnicity, is to increase logit P by .055. Similarly, the effect of being urban, relative to rural, controlling for education and ethnicity, is to increase logit P by .378. And the effect of being Indian, relative to Fijian, controlling for education and residence, is to increase logit P by 1.161.

The trouble is that logit P is not a familiar quantity, so that the meanings of these effects are not very clear.

To some extent, we can clarify the interpretation of these effects by writing logit P as $\log[P/(1-P)]$ and then examining the relationship between $P, P/(1-P)$, and $\log[P/(1-P)]$. We observe first that $P/(1-P)$ is a monotonically increasing function of P, and $\log[P/(1-P)]$ is a monotonically increasing function of $P/(1-P)$, as shown in Figure 5.4. [A function $f(P)$ is monotonically increasing if an increase in P always generates an increase in $f(P)$.] Figure 5.4 also shows the graph of $\log[P/(1-P)]$ against P.

Monotonicity means that if P increases, $P/(1-P)$ and $\log[P/(1-P)]$ also increase; if P decreases, $P/(1-P)$ and $\log[P/(1-P)]$ also decrease. Conversely, if $\log[P/(1-P)]$ increases, $P/(1-P)$ and P also increase; if $\log[P/(1-P)]$ decreases, $P/(1-P)$ and P also decrease.

In (5.21), a one-year increase in E generates an increase of .055 in $\log[P/(1-P)]$, which, because of monotonicity, goes hand in hand with increases in $P/(1-P)$ and P. *In other words, if b is positive, the effects of E on* $\log[P/(1-P)]$, $P/(1-P)$, *and P are positive. If b is negative, the effects of E on* $\log[P/(1-P)]$, $P/(1-P)$, *and P are all negative*. Similar statements can be made about the effects of U and I.

This is helpful, but still not very satisfactory. The next section further sharpens the conceptualization of "effect."

5.2.5. Odds Ratios as Measures of Effect on the Odds

From (5.20) and (5.16), our estimated model may be expressed as

$$\log \Omega = a + bE + cU + dI \tag{5.22}$$

Taking the exponential of both sides [i.e., taking each side of (5.22) as

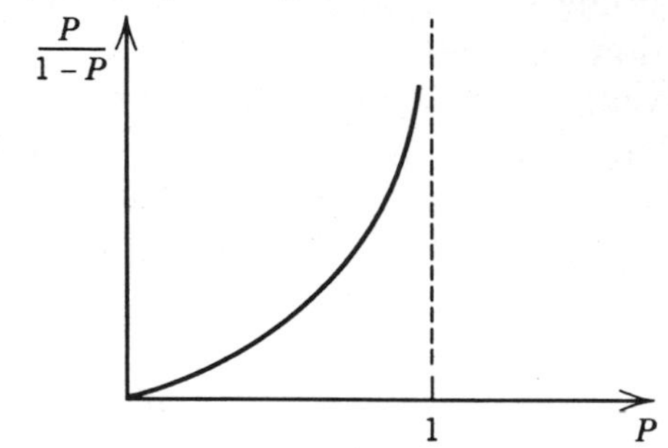

(b) $\log(P/(1 - P))$ plotted against $P/(1 - P)$

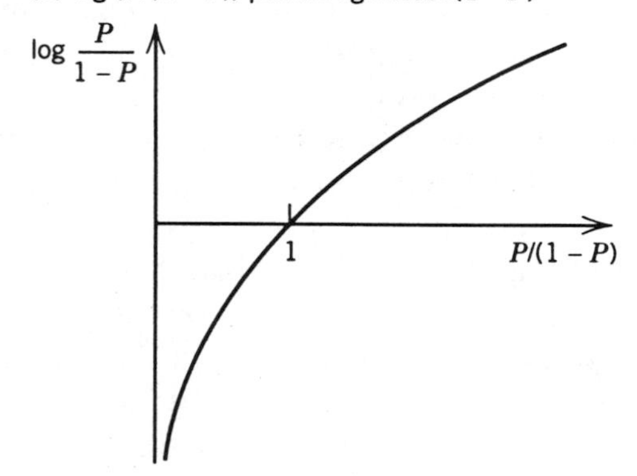

(c) $\log[P/(1 - P)]$ plotted against P

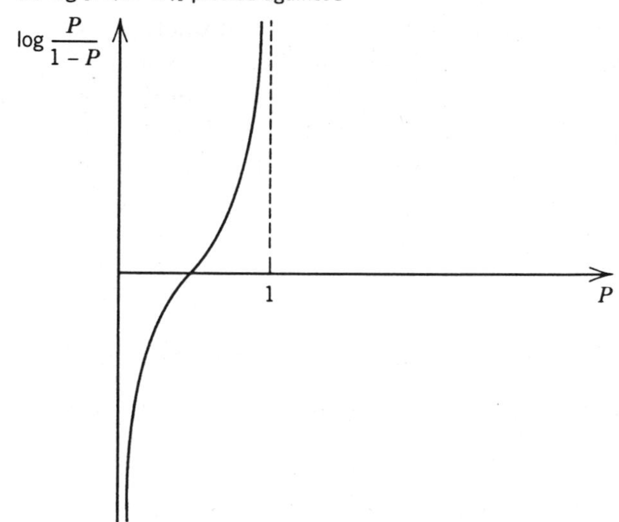

Figure 5.4. Graphic demonstration that $P/(1 - P)$ and $\log(P/(1 - P))$ are both monotonically increasing functions of P.

a power of e], we obtain

$$\Omega = e^{a+bE+cU+dI} \tag{5.23}$$

Suppose we increase E by one unit, holding U and I constant. Denoting the new value of Ω as Ω^*, we have

$$\begin{aligned} \Omega^* &= e^{a+b(E+1)+cU+dI} \\ &= e^{a+bE+cU+dI+b} \\ &= e^{a+bE+cU+dI}\, e^{b} \\ &= \Omega e^{b} \end{aligned} \tag{5.24}$$

which can be written alternatively as

$$\frac{\Omega^*}{\Omega} = e^{b} \tag{5.25}$$

From (5.24), it is evident that a one-unit increase in E, holding other predictor variables constant, multiplies the odds by the factor e^b. The quantity e^b is called an *odds ratio*, for reasons that are obvious from (5.25).

Equation (5.25) can be arrived at more directly by noting that, because (5.22) is in the form of a multiple regression equation, the effect of a one-unit increase in E, holding other predictor variables constant, is to increase $\log \Omega$ by b units. Thus

$$\log \Omega^* = \log \Omega + b \tag{5.26}$$

Taking each side of (5.26) as a power of e, we obtain

$$\Omega^* = \Omega e^{b} \tag{5.27}$$

which yields (5.25) when both sides are divided by Ω.

Let us review what has happened. *The original coefficient b represents the additive effect of a one-unit change in E on the log odds of using contraception. Equivalently, the odds ratio e^b represents the multiplicative effect of a one-unit change in E on the odds of using contraception*. Insofar as the odds is a more intuitively meaningful concept than the log odds, e^b is more readily understandable than b as a measure of effect.

The above discussion indicates that the logit model may be thought of as either an additive model or a multiplicative model, depending on how the response variable is conceptualized. *When we consider the log odds as the response variable, the logit model is an additive model, as in ordinary multiple regression. But when we consider the odds as the response variable, the logit model is a multiplicative model*, regarding the definition and interpretation of effects.

So far we have considered the effect of a one-unit change in E, which is a quantitative variable. What about a change in U, the dummy variable indicating urban or rural residence? If we consider the effect of a one-unit change in U (from 0 to 1) on the odds of using contraception, holding E and I constant, then, starting from (5.23) and using the same logic as before, we obtain

$$\Omega^* = \Omega e^c \tag{5.28}$$

In other words, the effect of being urban, relative to rural, controlling for education and ethnicity, is to multiply the odds of using contraception by e^c. Similarly, the effect of being Indian, relative to Fijian, controlling for education and residence, is to multiply the odds by e^d.

In our Fijian example [see equation (5.21) above], we have that $e^b = e^{.055} = 1.057$, $e^c = e^{.378} = 1.459$, and $e^d = e^{1.161} = 3.193$. Therefore, a one-year increase in E, holding U and I constant, multiplies the odds by 1.057 (equivalent to a 5.7 percent increase). A one-unit increase in U (from 0 to 1—i.e., from rural to urban), holding E and I constant, multiplies the odds by 1.459 (a 45.9 percent increase). A one-unit increase in I (from 0 to 1—i.e., from Fijian to Indian), holding E and U constant, multiplies the odds by 3.193 (a 219.3 percent increase).

5.2.6. The Effect on the Odds When the Predictor Variable Is Categorical with More Than Two Categories

Let us alter our model of the effect of education, residence, and ethnicity on contraceptive use, as follows:

P: estimated probability of contraceptive use
M: 1 if medium education, 0 otherwise
H: 1 if high education, 0 otherwise
U: 1 if urban, 0 otherwise
I: 1 if Indian, 0 otherwise

This is the same model as before, except that education is redefined as a categorical variable with three categories: low, medium, and high, with low as the reference category.

In log odds form, the model is

$$\log \Omega = a + bM + cH + dU + fI \tag{5.29}$$

From (5.29), we can calculate the following values of $\log \Omega$ for low, medium, and high education by setting (M, H) alternatively to $(0, 0)$, $(1, 0)$, and $(0, 1)$:

Low education	$(H = 0, M = 0)$:	$\log \Omega_L = a + dU + fI$
Medium education	$(H = 0, M = 1)$:	$\log \Omega_M = a + b + dU + fI$
High education	$(H = 1, M = 0)$:	$\log \Omega_H = a + c + dU + fI$

From these values of high, medium, and low education, we obtain

$$\begin{aligned} \log \Omega_M - \log \Omega_L &= (a + b + dU + fI) - (a + dU + fI) \\ &= b \end{aligned} \tag{5.30}$$

$$\begin{aligned} \log \Omega_H - \log \Omega_L &= (a + c + dU + fI) - (a + dU + fI) \\ &= c \end{aligned} \tag{5.31}$$

$$\begin{aligned} \log \Omega_M - \log \Omega_H &= (a + b + dU + fI) - (a + c + dU + fI) \\ &= b - c \end{aligned} \tag{5.32}$$

Equations (5.30)–(5.32) can be rewritten

$$\log \Omega_M = \log \Omega_L + b \tag{5.33}$$

$$\log \Omega_H = \log \Omega_L + c \tag{5.34}$$

$$\log \Omega_M = \log \Omega_H + (b - c) \tag{5.35}$$

Equation (5.33) says that the effect of medium education, relative to low education, controlling for U and I, is to increase $\log \Omega$ by b. Equation (5.34) says that the effect of high education, relative to low education, controlling for U and I, is to increase $\log \Omega$ by c. Equation (5.35) says that the effect of medium education, relative to high education, controlling for U and I is to increase $\log \Omega$ by $b - c$. It doesn't matter what values U and I are held constant at, because terms in U and I cancel in (5.30)–(5.32). Except for the form of the response variable, these results are the same as those obtained in Chapter 2 for ordinary multiple regression.

If equations (5.33)–(5.35) are taken as powers of e, these equations become

$$\Omega_M = \Omega_L e^b \tag{5.36}$$

$$\Omega_H = \Omega_L e^c \tag{5.37}$$

$$\Omega_M = \Omega_H e^{b-c} \tag{5.38}$$

Equation (5.36) says that the effect of medium education, relative to low education, controlling for U and I, is to multiply the odds by e^b. Equation (5.37) says that the effect of high education, relative to low education, controlling for U and I, is to multiply the odds by e^c. Equation (5.38) says that the effect of medium education, relative to high education, controlling for U and I, is to multiply the odds by e^{b-c}.

In equations (5.33)–(5.35), the effects on log Ω, which are additive, are b, c, and $b - c$. In equations (5.36)–(5.38), the effects on Ω, which are multiplicative, are e^b, e^c, and e^{b-c}.

Because (5.29) is in the form of a multiple regression equation, we can immediately say that the effect on logit P of, say, medium education, relative to high education, does not depend on which education category is chosen as the reference category. It follows that the effects on $P/(1 - P)$ and P itself do not depend on which category of education is chosen as the reference category. Changing the reference category does not require one to rerun the model. One can derive the education coefficients in the new model from the education coefficients in the original model. (See Section 2.4.2 of Chapter 2.) The coefficients of the other predictor variables are unaffected by changing the reference category for education.

5.2.7. The Effect of the Predictor Variables on the Risk P Itself

Let us return to our example of the effects of education (E), residence ($U = 1$ if urban, 0 otherwise), and ethnicity ($I = 1$ if Indian, 0 otherwise) on contraceptive use, where education is once again conceptualized as a quantitative variable, namely, the number of completed years of education.

If we wish to look at the effects of the predictor variables directly on the risk P, we must go back to the basic form of the logistic function:

$$P = \frac{1}{1 + e^{-(a+bE+cU+dI)}} \tag{5.39}$$

What is the effect on P of a one-unit increase in E, holding U and I constant? Denoting the new value of P as P^*, we get

$$P^* = \frac{1}{1 + e^{-(a+b(E+1)+cU+dI)}} \tag{5.40}$$

Both the difference, $P^* - P$ (representing the additive effect), and the ratio, P^*/P (representing the multiplicative effect), are functions not only of b but also of a, c, d, E, U, and I. Terms in a, c, d, E, U, and I do not cancel as they do in the case of the log odds (where all that remains in the difference is b) or the odds (where all the remains in the ratio is e^b). Therefore, both the additive effect and the multiplicative effect of a one-unit change in E on P depend on the levels of E, U, and I.

One way to handle this problem is to set U and I equal to their means in the sample and then to calculate and tabulate values of P for $E = 0, 1, 2, \ldots$. We shall return to this approach when we discuss how to present logit regression results in a multiple classification analysis (MCA) format.

Because the effects of the predictor variables on P are not as simple as the effects on Ω and $\log \Omega$, logit regression results are often presented as effects on Ω or $\log \Omega$.

5.2.8. Interactions

Let us now complicate the model considered in Section 5.2.7 by adding an interaction between residence and ethnicity. For reasons that will become clear shortly, it is convenient to represent logit P as $\log \Omega$. Our estimated model is then

$$\log \Omega = a + bE + cU + dI + fUI \tag{5.41}$$

Because (5.41) is in the form of an ordinary multiple regression equation, we can say immediately that the effect of a one-unit increase in U (from 0 to 1) on $\log \Omega$, holding E and I constant, is to change $\log \Omega$ by $c + fI$ (see Section 2.6 in Chapter 2). Denoting the new value of Ω as Ω^*, we therefore have that

$$\log \Omega^* = \log \Omega + (c + fI) \tag{5.42}$$

If we take each side of (5.42) as a power of e, (5.42) becomes

$$\Omega^* = \Omega e^{c+fI} \tag{5.43}$$

If f is positive, the multiplicative effect of a one-unit increase in U increases as I increases from 0 to 1 (i.e., from Fijian to Indian). If f is negative, the multiplicative effect of a one-unit increase in U decreases as I increases from 0 to 1.

It can similarly be shown that a one-unit increase in I, holding E and U constant, increases $\log \Omega$ by $d + fU$ and multiplies Ω by e^{d+fU}. If f is positive, the multiplicative effect of a one-unit increase in I increases as U increases from 0 to 1 (i.e., from rural to urban). If f is negative, the multiplicative effect of a one-unit increase in I decreases as U increases from 0 to 1.

Note that, in the context of (5.43), *interaction between U and I means that the multiplicative effect of U on* Ω *depends on the level of I. Similarly, the multiplicative effect of I on* Ω *depends on the level of U.*

5.2.9. Nonlinearities

Let us further complicate the model by adding a nonlinearity to the effect that E has on $\log \Omega$:

$$\log \Omega = a + bE + cE^2 + dU + fI + gUI \tag{5.44}$$

Again, because (5.44) is in the form of a multiple regression equation, we can say immediately that the (instantaneous) effect of a one-unit increase in E, holding U and I constant, is to change $\log \Omega$ by $b + 2cE$ (see Section 2.7.1 in Chapter 2). Denoting the new value of Ω as Ω^*, we therefore have that

$$\log \Omega^* = \log \Omega + (b + 2cE) \tag{5.45}$$

Taking each side of (5.45) as a power of e, we can rewrite (5.45) as

$$\Omega^* = \Omega e^{b+2cE} \tag{5.46}$$

If $c > 0$, the multiplicative effect of a one-unit increase in E on Ω increases as E increases. If $c < 0$, the multiplicative effect of a one-unit increase in E on Ω decreases as E increases.

Note that, in the context of (5.46), *nonlinearity in E means that the multiplicative effect of E on* Ω *depends on the level of E.*

5.3. STATISTICAL INFERENCE

5.3.1. Tests of Coefficients

The computer output for logit regression includes estimated standard errors and p values for the coefficients. The standard error of a coefficient, b, is interpreted as the standard error of the sampling distribution of b, just as in ordinary regression. As before in this book, we shall accept the standard errors on faith and not be concerned about the details of the statistical theory underlying their derivation.

From a coefficient b and its standard error s_b, a Z *value* can be calculated as

$$\boxed{Z \equiv \frac{b - \beta}{s_b}} \tag{5.47}$$

where β is the hypothesized value of the coefficient in the underlying population. Usually our null hypothesis is that $\beta = 0$. then (5.47) reduces to

$$Z \equiv \frac{b}{s_b} \tag{5.48}$$

If our hypothesized value of the coefficient in the underlying population is correct, then, for large samples, the sampling distribution of Z is approximately standard normal with mean 0 and variance 1. How large is large? A rule of thumb that is sometimes used is the following: As long as the ratio of sample size to the number of parameters to be estimated is at least 20, the normal approximation is usually satisfactory. The sampling distribution of Z for small samples is more complicated and is not considered in this book.

Given the coefficients and their standard errors and Z values, hypothesis testing and the construction of confidence intervals are done in the same way as in ordinary multiple regression, using the table of standard normal probabilities (Table B2, Appendix B). *One does not use the table of t critical points in Table B.1 (Appendix B).* In Chapter 2 on multiple regression, the formulae analogous to (5.47) and (5.48) had t in place of Z. In logit regression, however, the sampling distribution of $(b - \beta)/s_b$ does not conform to Student's t distribution.

In logit regression as in ordinary multiple regression, one can optionally print out a regression coefficient covariance matrix and test whether two coefficients (e.g., the coefficients of M and H, denoting medium and high education) differ significantly from each other. The test procedure is the same as that described earlier in Section 2.9.2 of Chapter 2.

If one is dealing with the odds instead of the log odds, then one needs to construct confidence intervals for e^b instead of b. If the boundaries of the confidence interval for b are u and v, then the boundaries for the confidence interval for e^b are calculated as e^u and e^v.

5.3.2. Testing the Difference Between Two Models

As we shall discuss in more detail later, the *likelihood* L is the probability of observing our particular sample data under the assumption that the fitted model is true. Thus L is somewhere between 0 and 1. The *log likelihood, or* $\log L$ (log to the base e), is often mathematically more convenient to work with than L itself. *Because* L *lies between* 0 *and* 1, $\log L$ *is negative.*

Suppose that we have two logit regression models which have the same response variable but different sets of predictor variables, where the second model has all the predictor variables included in the first, plus at least one more. We say that the first model is *nested* in the second model.

Let us denote the likelihood of the first model by L_1, and the likelihood of the second model by L_2. One way to test whether the two models differ significantly from each other might be to test whether L_1 differs significantly from L_2, or, equivalently, whether $L_2 - L_1$ differs significantly from zero. But his is not possible, because the sampling distribution of $L_2 - L_1$ is not known.

Alternatively one might think of testing whether L_1/L_2 differs significantly from 1. But the sampling distribution of L_1/L_2 is not known either.

Fortunately, a statistic for which the sampling distribution is known is $-\log(L_1/L_2)^2$, where $L_1 < L_2$. This can be written as

$$\begin{aligned} -\log(L_1/L_2)^2 &= -2\log(L_1/L_2) \\ &= -2(\log L_1 - \log L_2) \\ &= (-2\log L_1) - (-2\log L_2) \end{aligned} \tag{5.49}$$

The quantity L_1/L_2 is called the *likelihood ratio*. The condition $L_1 < L_2$, which is not restrictive, simply means that the first model is nested in the second model. When $L_1 < L_2, -\log(L_1/L_2)^2$ is positive.

We can use (5.49) *to test whether the second model fits the data significantly better than the first model. The test is a simple* χ^2 *test. It turns out that* $-\log(L_1/L_2)^2$ (*i.e., the positive difference in* $-2\log L$ *between the two models*) *is distributed approximately as* χ^2 *with degrees of freedom equal to the difference in the number of coefficients to be estimated in the two models*. The word "distributed" refers here to the sampling distribution of the difference in $-2\log L$; if we took repeated samples from the same underlying population, each time fitting the two models and computing the difference in $-2\log L$ between the two fitted models, we would find that the difference in $-2\log L$ would be distributed as just mentioned.

For example, suppose as a real-life example, for fecund, nonpregnant, currently married women aged 35–44 in the 1974 Fiji Fertility Survey, that the two fitted models are

$$\text{logit } P = .363 - .034E + .597U \tag{5.50}$$

and

$$\text{logit } P = -.611 + .055E + .378U + 1.161I \tag{5.51}$$

where P denotes the probability of using contraception, E denotes number of completed years of education, U denotes residence (1 if urban, 0 otherwise), and I denotes ethnicity (1 if Indian, 0 otherwise).

The $-2\log L$ statistic is 1264.6 for the first model and 1213.1 for the second model. (These values are automatically printed out by the SAS program for binary logit regression.) The difference in $-2\log L$ between the two models is therefore 51.5. The difference in the number of coefficients to be estimated is 1, so d.f. = 1. Consulting the χ^2 table in Appendix B (Table B.4), we find that χ^2 with d.f. = 1 is 10.8 for $p = .001$, which is the highest level of significance included in the table. Because our observed value of χ^2 is 51.5, which is greater than the critical point of 10.8, we can say that the two models differ with an observed level of significance of $p < .001$. We conclude that the model in (5.51) fits the data much better than the model in (5.50).

In (5.50) and (5.51), $-2\log L$ is approximately 1200, which means that $\log L$ is approximately -600 and L is approximately e^{-600}, which is approximately 10^{-261}. This extremely small number, representing the likelihood of the observed data under the assumption that the model is true, might lead one to think that the model fits poorly. It must be remembered, however, that each individual observation has a probability, and that the overall likelihood is the product of these individual probabilities. Each individual probability is a number less than one. If the sample size is large, the product of the individual probabilities will be an extremely small number, even if the individual probabilities are fairly close to one. Therefore, a very small likelihood does not necessarily mean a poor fit.

Very frequently, the two models compared are the model in which we are interested, which we may refer to as the *test model*, and what is often called the *intercept model*. For example, if the test model is specified by (5.51), the corresponding intercept model is

$$\text{logit } P = a \tag{5.52}$$

where a is a constant to be fitted. In effect, we ask whether the test model differs significantly from a baseline model with no predictor variables. Computer programs for logit regression typically print out not only the value of $-2 \log L$ for both the test model and the intercept model, but also the p value for the difference in $-2 \log L$ between these two models. In this case, there is no need to consult a χ^2 table, because the computer does it for us. For the Fiji data, the logit regression output for the test model in (5.51) indicates a p value of .000 for the difference between this model and the intercept model. Because .000 is a rounded value, this means that $p < .0005$. In other words, the test model fits the data much better than the intercept model.

5.4. GOODNESS OF FIT

In multiple regression, the traditional indicator of goodness of fit is R^2, which measures the proportion of variation in the response variable that is explained by the predictor variables. In the case of logit regression, one could also calculate the proportion of variation in the response variable that is explained by the predictor variables, but in this case it is impossible for the observed values of the response variable, which are either 0 or 1, to conform exactly to the fitted values of P. The maximum value of this proportion depends on the mean and variance of P (Morrison, 1972).

An alternative measure of goodness of fit may be derived from the likelihood statistic. Let L_0 denote the likelihood for the fitted intercept model, and let L_1 denote the likelihood for the fitted test model. Define pseudo-R^2 as

$$\text{pseudo-}R^2 = \frac{1 - (L_0/L_1)^{2/n}}{1 - L_0^{2/n}} \tag{5.53}$$

$$= \frac{L_1^{2/n} - L_0^{2/n}}{1 - L_0^{2/n}} \tag{5.54}$$

where n is the sample size. The minimum value of this quantity is zero when the fit is as bad as it can be (when $L_1 = L_0$), and the maximum value is one when the fit is as good as it can be (when $L_1 = 1$). This definition of pseudo-R^2 was suggested by Cragg and Uhler (1970). Unfortunately there is no formal significance test that utilizes this measure.

Another definition suggested by McFadden (1974) is simply

$$\text{pseudo-}R^2 = 1 - (\log L_0/\log L_1)^{2/n} \tag{5.55}$$

There is no formal test that utilizes this measure either.

The definition of pseudo-R^2 used by the SAS program for logit regression is

$$\boxed{\text{pseudo-}R^2 = \frac{2\log L_1 - 2\log L_0 - 2k}{-2\log L_0}} \tag{5.56}$$

where k denotes the number of coefficients to be estimated, not counting the intercept. Recall from the previous section that the quantity $-2\log L_0 - 2\log L_1$ is model chi-square. If model chi-square is less than $2k$ in (5.56), pseudo-R^2 is set to zero (Harrell, 1983). Pseudo-R, computed as the square root of pseudo-R^2, is sometimes presented instead of pseudo-R^2. Pseudo-R, computed for the model presented earlier in equation (5.21), is .22.

There are several difficulties with these measures of pseudo-R^2, some of which have already been mentioned. First, there are several different measures available, which can give rather different numerical results when applied to the same data set. Second, there is little basis for choosing one measure over the other. Third, statistical tests that utilize pseudo-R^2 are not available for any of the measures. For these reasons, many authors do not present values of pseudo-R^2 when reporting results from logit regression analyses. If the analyst does use one of the measures, he or she should report which one.

5.5. MCA ADAPTED TO LOGIT REGRESSION

When multiple classification analysis (MCA) is adapted to logit regression, both unadjusted and adjusted values of the response variable can be calculated, just as in ordinary MCA. The unadjusted values are based on logit regressions that incorporate one predictor variable at a time, and the adjusted values are based on the complete model including all predictor variables simultaneously. We shall consider only adjusted values in this chapter.

Although pseudo-R is the analogue of ordinary R, there is no logit regression analogue to partial R. Therefore, measures of partial association are omitted from the MCA tables based on logit regression.

Statistical computing packages such as SPSS, SAS, BMDP, and LIMDEP do not include MCA programs for logit regression. The analyst must construct the MCA tables from the underlying logit regressions, as described below.

5.5.1. An Illustrative Example

To illustrate how MCA can be adapted to logit regression, we consider the contraceptive use effect of age, age-squared, education (low, medium, and high), residence (urban, rural), ethnicity (Indians, Fijians), and residence × ethnicity. Whereas we previously considered fecund, nonpregnant, currently married women aged 35–44, we now broaden the age range to 15–49 while at the same time introducing a control for age and age-squared. An age-squared term is included because contraceptive use tends to increase with age up to about age 35 or 40, after which it tends to level off or even decline. This kind of simple curvature is readily captured by a quadratic term, as discussed earlier in Chapter 2.

Our variables are

P: estimated probability of contraceptive use

A: age

M: 1 if medium education, 0 otherwise
H: 1 if high education, 0 otherwise
U: 1 if urban, 0 otherwise
I: 1 if Indian, 0 otherwise

In log odds form, the model is

$$\log \Omega = a + bA + cA^2 + dM + fH + gU + hI + jUI \quad (5.57)$$

In odds form, the model is

$$\Omega = e^{a+bA+cA^2+dM+fH+gU+hI+jUI} \quad (5.58)$$

In probability form, the model is

$$P = \frac{1}{1 + e^{-(a+bA+cA^2+dM+fH+gU+hI+jUI)}} \quad (5.59)$$

When the model is fitted to the Fiji data, it is found, as shown in Table 5.1, that all coefficients except j (the coefficient of the UI

TABLE 5.1. Coefficients and Effects (Odds Ratios) for the Estimated Model, $\log \Omega = a + bA + cA^2 + dM + fH + gU + hI + jUI$: Fecund, Nonpregnant, Currently Married Women Aged 15–49 in the 1974 Fiji Fertility Survey[a]

Predictor Variable	Coefficient (S.E.)	Effect (Odds Ratio)
Intercept	−5.594*(.610)	
Age		$e^{.255+2(-.003)A} = 1.290\, e^{-.006A}$
A	.255*(.038)	—
A^2	−.003*(.001)	—
Education		
H	.331*(.102)	$e^{.331} = 1.392$
M	.254*(.096)	$e^{.254} = 1.289$
Residence × ethnicity		
U	.294*(.127)	$e^{.294-.046I} = 1.342\, e^{-.046I}$
I	1.160*(.099)	$e^{1.160-.046U} = 3.190\, e^{-.046U}$
UI	−.046 (.158)	—

[a]The response variable is the log odds of contraceptive use. An asterisk after a coefficient indicates that the coefficient differs from zero with a two-sided $p < .05$. Numbers in parentheses following coefficients are standard errors of coefficients. The effect of age depends on the level of A, the effect of residence depends on the level of I, and the effect of ethnicity depends on the level of U; if A, U, and I are set to their mean values of 31.03, .353, and .609, respectively, these effects are 1.071 for age, 1.305 for residence (U), and 3.139 for ethnicity (I).

interaction) are significant at the 5 percent level or higher. Although the *UI* term could be dropped from the model (the model would then have to be reestimated without the *UI* term), we retain it for purposes of illustration. Table 5.1 also shows effects, as measured by odds ratios.

5.5.2. Table Set-up and Numerical Results

Table 5.2 shows the MCA table set-up and numerical results for adjusted values of log Ω. The model is given by equation (5.57), which is repeated for convenience in the footnote to the table. Because (5.57) is in the form of a multiple regression equation, the table set-up is done in exactly the same way as in ordinary multiple regression, as described in Chapter 3. Although unadjusted values are not shown here, they would also be set up as in Chapter 3, had we included them.

Because age is a continuous variable, one must enter selected values of age, which play the role of categories. We have chosen 15, 25, 35, and 45. All the other variables are categorical to begin with and require no selection of values. In each panel, other variables are controlled by setting these variables at their mean values in the entire sample, and these mean values are given in the footnote to the table. When the values of the coefficients and mean values are substituted into the formulae in the first column of the table, one obtains the numerical values given in the second column of the table. Using the information given in the footnote to the table, the reader can replicate the numerical values in the second column of the table as a check on understanding.

Whereas Table 5.2 gives adjusted values of log Ω, Table 5.3 gives corresponding adjusted values of Ω and P. The adjusted values of Ω are obtained by taking the adjusted values of log Ω in Table 5.2 as a power of e, that is, $\Omega = e^{\log \Omega}$. Adjusted values of P are then calculated as

$$P = \frac{\Omega}{1 + \Omega} \tag{5.60}$$

This formula is derived by solving the equation $\Omega \equiv P/(1 - P)$ for P.

The numerical results in Tables 5.2 and 5.3 show that there is indeed a tendency for contraceptive use to level off at the older reproductive ages, which justifies the inclusion of the age-squared term in the model. As mentioned earlier, the interaction between

TABLE 5.2. MCA Table of Adjusted Values of the Log Odds of Contraceptive Use for the Estimated Model, $\log \Omega = a + bA + cA^2 + dM + fH + gU + hI + jUI$: Fecund, Nonpregnant, Currently Married Women Aged 15–49 in the 1974 Fiji Fertility Survey[a–c]

Predictor Variable	n	Adjusted Value of $\log \Omega$: Formula	Numerical Value
Age			
15	—	$a + 15b + (15)^2c + d\overline{M} + f\overline{H} + g\overline{U} + h\overline{I} + j\overline{UI}$	-1.491
25	—	$a + 25b + (25)^2c + d\overline{M} + f\overline{H} + g\overline{U} + h\overline{I} + j\overline{UI}$	$-.204$
35	—	$a + 35b + (35)^2c + d\overline{M} + f\overline{H} + g\overline{U} + h\overline{I} + j\overline{UI}$	.451
45	—	$a + 45b + (45)^2c + d\overline{M} + f\overline{H} + g\overline{U} + h\overline{I} + j\overline{UI}$	.473
Education			
Low	1266	$a + b\overline{A} + c\overline{A^2} + g\overline{U} + h\overline{I} + j\overline{UI}$	$-.106$
Medium	1044	$a + d + b\overline{A} + c\overline{A^2} + g\overline{U} + h\overline{I} + j\overline{UI}$	.148
High	1169	$a + f + b\overline{A} + c\overline{A^2} + g\overline{U} + h\overline{I} + j\overline{UI}$	.226
Residence and ethnicity			
Urban Indian	854	$a + g + h + j + b\overline{A} + c\overline{A^2} + d\overline{M} + f\overline{H}$	.691
Urban Fijian	375	$a + g + b\overline{A} + c\overline{A^2} + d\overline{M} + f\overline{H}$	$-.423$
Rural Indian	1265	$a + h + b\overline{A} + c\overline{A^2} + d\overline{M} + f\overline{H}$	.443
Rural Fijian	985	$a + b\overline{A} + c\overline{A^2} + d\overline{M} + f\overline{H}$	$-.717$
Total	3479		

[a]The estimated model is

$$\log \Omega = -5.594 + .255A - .003A^2 + .254M + .331H + .294U + 1.160I - .046UI$$

Mean values of the predictor variables are $\overline{A} = 31.03$, $\overline{A^2} = 1021.66$, $\overline{M} = .300$, $\overline{H} = .336$, $\overline{U} = .353$, $\overline{I} = .609$, and $\overline{UI} = .245$. Note that A^2 and UI are treated as separate variables, so that means are calculated as the mean of A^2 (instead of the square of the mean of A) and the mean of UI (instead of the mean of U times the mean of I). $-2 \log L_0 = 4817.20$ for the intercept model and $-2 \log L_1 = 4462.12$ for the test model, given above. The test model differs significantly from the intercept model, with $p < .001$ from the likelihood ratio test. Pseudo-R for the test model is .27.

[b]The values of $\log \Omega$ in the table were actually calculated from values of coefficients and means accurate to seven significant figures. It turns out that, in this example, the use of figures rounded to only three decimal places introduces substantial errors in the estimates of $\log \Omega$, Ω, and P. The main source of these errors is the use of $-.003$ instead of $-.003162000$ for the coefficient of A^2.

[c]To be on the safe side, intermediate calculations should always employ data values with at least two more significant figures than the number of significant figures desired in the final estimate. Note that significant figures are not the same as decimal places. The data value .002 has three decimal places but only one significant figure, because leading zeros are not counted as significant figures. The data value 1.003 has three decimal places but four significant figures. The number 1.0030 has five significant figures, because the last zero indicates that the data value is accurate to four decimal places.

TABLE 5.3. MCA Table of the Adjusted Effects of Age, Education, Residence, and Ethnicity on the Odds of Contraceptive Use and the Probability of Contraceptive Use for the Estimated Model, $\log \Omega = a + bA + cA^2 + dM + fM + gU + hI + jUI$: Fecund, Nonpregnant, Currently Married Women Aged 15–49 in the 1974 Fiji Fertility Survey[a]

Predictor Variable	n	Adjusted Value Ω	Adjusted Value P
Age			
15	—	.225	.18
25	—	.815	.45
35	—	1.569	.61
45	—	1.605	.62
Education			
Low	1266	.900	.47
Medium	1044	1.160	.54
High	1169	1.253	.56
Residence and ethnicity			
Urban Indian	854	1.995	.67
Urban Fijian	375	.655	.40
Rural Indian	1265	1.557	.61
Rural Fijian	985	.488	.33
Total	3479		

[a]Values of Ω are calculated by taking the values of $\log \Omega$ in Table 5.2 as a power of e. Values of P are then calculated as $\Omega/(1 + \Omega)$. Among the 3479 women, 1810 are users and 1669 are nonusers. Values of Ω and P were calculated from more accurate values of $\log \Omega$ than shown in Table 5.2; see footnote in Table 5.2.

residence and ethnicity is not statistically significant, although we have left the interaction in the model. Urban women have somewhat higher rates of contraceptive use than do rural women, controlling for the other variables in the model. Indian women have considerably higher rates of contraceptive use than do Fijian women, controlling for the other variables in the model.

Although the probabilities in the last column of Table 5.3 are presented in decimal form, they are frequently presented in the literature in percentage form. In percentage form, the value of P for, say, urban Indians would be presented as 67 percent instead of .67.

Normally, in a published research paper in an academic journal, one would not present formulae for adjusted $\log \Omega$, as shown in Table 5.2. Instead, one would omit $\log \Omega$ altogether from the tables and combine Tables 5.2 and 5.3 into a single table, with columns for numerical values of n, adjusted Ω, and adjusted P. Because adjusted Ω can be calculated from adjusted P as $\Omega = P(1 - P)$, even adjusted Ω would usually be omitted.

5.6. FITTING THE LOGIT REGRESSION MODEL

To fit the logit regression model, we use the *method of maximum likelihood*.

5.6.1. The Maximum Likelihood Method

To illustrate this method, let us consider a very simple model:

$$\text{logit } P = a + bX \tag{5.61}$$

which can be written alternatively as

$$P = \frac{1}{1 + e^{-(a+bX)}} \tag{5.62}$$

We assume that the mathematical form of (5.61) and (5.62) is correct, but we don't yet know the values of a and b, which are treated as unknowns.

The first step is to formulate a *likelihood function*, L. As mentioned earlier, L is the probability of observing our particular sample data under the assumption that the model is true. That is, we assume that the mathematical form of the model, as given by (5.61) or (5.62), is correct, but we don't yet know the values of a and b. *We choose a and b so that L is maximized. In other words, we choose values of the unknown parameters that maximize the likelihood of the observed data and call these the best-fitting parameters.* The method can be thought of as considering all possible combinations of a and b, calculating L for each combination, and picking the combination that yields the largest value of L. In practice we use a mathematical shortcut, which is described below.

5.6.2. Derivation of the Likelihood Function

Although we shall not usually be concerned about the mathematical details of maximum likelihood estimation, it is nevertheless useful to examine a simple example to provide a better feel for how the method works.

Suppose that our original response variable is Y, which for individuals is either 0 or 1. Our model is

$$P = \frac{1}{1 + e^{-(a+bX)}} \qquad \text{(5.62 repeated)}$$

Denote the first individual by subscript 1. For this individual $(X, Y) = (X_1, Y_1)$. We then have that

$$P_1 = \frac{1}{1 + e^{-(a+bX_1)}} \qquad (5.63)$$

Similarly, for individual 2,

$$P_2 = \frac{1}{1 + e^{-(a+bX_2)}} \qquad (5.64)$$

And so on for the remaining individuals in the sample.

Let us consider individual 1 in more detail. We have from (5.63) that

$$\Pr(Y_1 = 1) = P_1 \qquad (5.65)$$

where Pr denotes "probability that." Therefore,

$$\Pr(Y_1 = 0) = 1 - P_1 \qquad (5.66)$$

We can combine (5.65) and (5.66) into one formula:

$$\Pr(Y_1) = P_1^{Y_1}(1 - P_1)^{(1-Y_1)} \qquad (5.67)$$

Let us check that this equation works. If $Y_1 = 1$, then

$$\Pr(Y_1 = 1) = P_1^1(1 - P_1)^{(1-1)} = P_1 \qquad \textit{Checks} \qquad (5.68)$$

If $Y_1 = 0$, then

$$\Pr(Y_1 = 0) = P_1^0(1 - P_1)^{(1-0)} = 1 - P_1 \qquad \textit{Checks} \qquad (5.69)$$

Similarly

$$\Pr(Y_2) = P_2^{Y_2}(1 - P_2)^{(1-Y_2)} \qquad (5.70)$$

And so on up to $\Pr(Y_n)$, where n denotes the number of sample cases.

If we assume simple random sampling, these n probabilities are independent. Then the probability, or likelihood, of observing our particular sample data is

$$
\begin{aligned}
\Pr(Y_1, Y_2, \ldots, Y_n) &= \Pr(Y_1)\Pr(Y_2)\ldots\Pr(Y_n) \\
&= \prod_{i=1}^{n} \Pr(Y_i) \\
&= \prod \left[P_i^{Y_i}(1 - P_i)^{(1-Y_i)} \right] \\
&= \prod \left[\frac{1}{1 + e^{-(a+bX_i)}} \right]^{Y_i} \left[1 - \frac{1}{1 + e^{-(a+bX_i)}} \right]^{1-Y_i} \\
&= L(a, b) \qquad (5.71)
\end{aligned}
$$

where $\prod$ is the product symbol (analogous to the summation symbol Σ), and where $L(a, b)$ indicates that the likelihood function, L, is a function of the unknown parameters, a and b. Note that the X_i and Y_i are observed data and therefore constants in the equation, not unknown parameters.

It is also useful to derive a formula for log L:

$$
\begin{aligned}
\log L(a, b) = \sum \Bigg\{ Y_i \log\left[\frac{1}{1 + e^{-(a+bX_i)}} \right] \\
+ (1 - Y_i)\log\left[1 - \frac{1}{1 + e^{-(a+bX_i)}} \right] \Bigg\} \qquad (5.72)
\end{aligned}
$$

In getting from (5.71) to (5.72), we make use of the rules $\log xy = \log x + \log y$ and $\log x^y = y \log x$.

We wish to find the values of a and b that maximize $L(a, b)$. Because log L is a monotonically increasing function of L, maximizing L is equivalent to maximizing log L. In other words, the same values of a and b that maximize L also maximize log L. Mathematically, it is easier to maximize log L than to maximize L directly. We maximize log L by taking the partial derivative of log L first with respect to a and second with respect to b, and then equating each derivative separately to zero, yielding two equations in two unknowns, a and b. These equations are then solved for a and b by numerical methods that are beyond the scope of this book. Derivation of the standard errors of a and b is also beyond the scope of this book.

We have sketched the essentials of this application of the maximum likelihood estimation (MLE) method in order to give the reader a feel for how the method works. MLE is also used to fit the multinomial logit model and the hazard model, which are considered in later chapters. Henceforth, we shall simply pronounce the magic words "maximum likelihood" and accept on faith that the model has been fitted. We shall then focus on model specification and interpretation, as in earlier chapters.

5.7. SOME LIMITATIONS OF THE LOGIT REGRESSION MODEL

In logit regression, the lower and upper asymptotes of the logistic function for P are 0 and 1. In some applications, this may not be realistic. Suppose, for example, that we are looking at the effect of education on contraceptive use in the United States. In the United States, even persons with no education have a probability of using contraception that is well above zero, so that a lower asymptote of zero may be unrealistic. This is not serious a problem as it might seem, however, because when we fit the logit regression model, we do not use the entire curve. In effect, we use only a piece of the curve corresponding to the observed range of the predictor variable.

5.8. FURTHER READING

A simple discussion of logit regression may be found in Wonnacott and Wonnacott (1979), *Econometrics*. A somewhat more advanced treatment is found in Hanushek and Jackson (1977), *Statistical Methods for Social Scientists* and in Fox (1984), *Linear Statistical Models and Related Methods*. For more advanced treatments, see Walker and Duncan (1967), Cox (1970), Amemiya (1981), and Maddala (1983).

6

MULTINOMIAL LOGIT REGRESSION

The *multinomial logit model* (also called the *polytomous logit model*) is a generalization of the binary logit model considered in Chapter 5. In this context, "binary" means that the response variable has two categories, and "multinomial" means that the response variable has three or more categories. As in binary logit regression, the predictors in multinomial logit regression may be quantitative, categorical, or a mixture of the two.

6.1. FROM LOGIT TO MULTINOMIAL LOGIT

6.1.1. The Basic Form of the Multinomial Logit Model

The explication of the multinomial logit model is facilitated by a simple example. Suppose that the response variable is contraceptive method choice:

P_1: estimated probability of using sterilization
P_2: estimated probability of using some other method
P_3: estimated probability of using no method

The categories of the response variable are mutually exclusive and exhaustive: A sample member must fall in one and only one of the categories.

Suppose that the reference category is "no method." As in previous chapters, the choice of reference category is arbitrary, insofar as this choice has no effect on the final estimated probabilities of using each method.

Suppose also that the predictor variables are education (low, medium, or high) and ethnicity (Indian or Fijian):

M: 1 if medium education, 0 otherwise
H: 1 if high education, 0 otherwise
I: 1 if Indian, 0 otherwise

Our theory is that education and ethnicity influence contraceptive method choice.

The multinomial logit model then consists of two equations plus a constraint:

$$\log \frac{P_1}{P_3} = a_1 + b_1 M + c_1 H + d_1 I \tag{6.1a}$$

$$\log \frac{P_2}{P_3} = a_2 + b_2 M + c_2 H + d_2 I \tag{6.1b}$$

$$P_1 + P_2 + P_3 = 1 \tag{6.1c}$$

Logarithms are natural logarithms (base e). In general, the number of model equations (including the constraint) equals the number of categories of the response variable.

Strictly speaking, the quantities P_1/P_3 and P_2/P_3 in (6.1) are not odds, because numerator and denominator do not necessarily sum to one. We may think of P_1/P_3 and P_2/P_3 as "improper" odds. For convenience, however, and in accordance with common usage, we shall refer to them simply as odds. Each of these odds has for its denominator the probability of the reference category of the response variable.

The model in (6.1) can be fitted by the method of maximum likelihood. We assume that the mathematical form of the model is correct, and we choose the values of a_1, b_1, c_1, d_1, a_2, b_2, c_2, and d_2 to maximize the likelihood function. We shall not worry about the mathematical details of how this is done. Instead, we shall assume that the model has been fitted and proceed to examine how to use and interpret it.

6.1.2. Interpretation of Coefficients

In multinomial logit regression, the interpretation of coefficients is not as straightforward as in binary logit regression.

Suppose, for example, that d_1 is positive in (6.1a). Then a one-unit increase in I (from 0 to 1) causes $\log(P_1/P_3)$ to increase by d_1 units. When $\log(P_1/P_3)$ increases, the odds P_1/P_3 also increase, because $\log(P_1/P_3)$ is a monotonic-increasing function of P_1/P_3.

However, we cannot reason that P_1 itself increases. P_1 could actually decrease. This could happen if P_3 also decreases and if the proportionate decrease in P_3 exceeds the proportionate decrease in P_1. In sum, P_1/P_3 can increase while both P_1 and P_3 decrease.

Therefore, a positive value of d_1 does not necessarily mean that a one-unit increase in I acts to increase P_1. The opposite may be true.

This could not happen in binary logit regression, because the numerator and denominator of $P/(1-P)$ cannot move in the same direction. If P increases, $1-P$ must decrease by the same amount. Therefore, if $P/(1-P)$ increases, P must also increase.

The above discussion illustrates that, in multinomial logit regression, the effects of the predictor variables on $\log(P_1/P_3)$ and P_1/P_3 can be misleading, because the effects on P_1 can be in the opposite direction. The same point applies to the effects of the predictor variables on $\log(P_2/P_3)$, P_2/P_3, and P_2. Therefore, in presenting results of multinomial logit analysis, we deemphasize the odds and log odds and focus instead on the effects of the predictor variables directly on P_1, P_2, and P_3.

6.1.3. Presentation of Results in Multiple Classification Analysis (MCA) Format

As in binary logit regression, the most convenient way to present the effects of the predictor variables on P_1, P_2, and P_3 is in the form of an MCA table, which is constructed in the following way:

The first step is to take each side of (6.1a) and (6.1b) as a power of e and then multiply through by P_3, yielding

$$P_1 = P_3 e^{a_1 + b_1 M + c_1 H + d_1 I} \tag{6.2a}$$

$$P_2 = P_3 e^{a_2 + b_2 M + c_2 H + d_2 I} \tag{6.2b}$$

We also have the identity

$$P_3 = P_3 \tag{6.2c}$$

If we now add (6.2a), (6.2b), and (6.2c) and recall that $P_1 + P_2 + P_3 = 1$, we get

$$1 = P_3 \sum_{j=1}^{2} e^{a_j + b_j M + c_j H + d_j I} + P_3 \tag{6.3}$$

Solving (6.3) for P_3, we obtain

$$P_3 = \frac{1}{1 + \sum e^{a_j + b_j M + c_j H + d_j I}} \tag{6.4}$$

where the notation for the summation is now abbreviated to omit the limits of summation.

Substituting (6.4) back into (6.2a) and (6.2b) and repeating (6.4), we obtain

$$P_1 = \frac{e^{a_1 + b_1 M + c_1 H + d_1 I}}{1 + \sum e^{a_j + b_j M + c_j H + d_j I}} \tag{6.5a}$$

$$P_2 = \frac{e^{a_2 + b_2 M + c_2 H + d_2 I}}{1 + \sum e^{a_j + b_j M + c_j H + d_j I}} \tag{6.5b}$$

$$P_3 = \frac{1}{1 + \sum e^{a_j + b_j M + c_j H + d_j I}} \tag{6.5c}$$

where the summations range from $j = 1$ to $j = 2$. Equations (6.5), which are an alternative statement of the model in equations (6.1), are calculation formulae for P_1, P_2, and P_3. A shortcut for calculating P_3 is to calculate P_1 and P_2 from (6.5a) and (6.5b) and then to obtain P_3 as $1 - (P_1 + P_2)$.

The MCA table is constructed by substituting appropriate combinations of ones, zeros, and mean values in equations (6.5), as shown in Table 6.1. For example, the formulae for P_1, P_2, and P_3 for those with high education in Table 6.1 are obtained by substituting $M = 0$, $H = 1$, and $I = \bar{I}$ in (6.5a), (6.5b), and (6.5c). The formulae for P_1, P_2, and P_3 for Fijians are obtained by substituting $I = 0$, $M = \bar{M}$, and $H = \bar{H}$ in (6.5a), (6.5b), and (6.5c).

The calculation of the MCA table for multinomial logit regression can be very tedious on a desk calculator. The calculation is most easily accomplished by transferring the multinomial logit computer output (fitted coefficients along with the mean values of the predictor variables) into a spreadsheet program like LOTUS.

TABLE 6.1. Set-up for MCA Table of Adjusted Values of P_j for the Model $\log P_j / P_3 = a_j + b_j M + c_j H + d_j I$, $j = 1, 2$: Fecund, Nonpregnant, Currently Married Women Aged 35–44 in the 1974 Fiji Fertility Survey [a]

Predictor Variable	n	P_1 (Sterilization)	P_2 (Other Methods)	P_3 (No Method)
Education				
Low	n_L	$\dfrac{e^{a_1+d_1\bar{I}}}{1+\Sigma e^{a_j+d_j\bar{I}}}$	$\dfrac{e^{a_2+d_2\bar{I}}}{1+\Sigma e^{a_j+d_j\bar{I}}}$	$\dfrac{1}{1+\Sigma e^{a_j+d_j\bar{I}}}$
Medium	n_M	$\dfrac{e^{a_1+b_1+d_1\bar{I}}}{1+\Sigma e^{a_j+b_j+d_j\bar{I}}}$	$\dfrac{e^{a_2+b_2+d_2\bar{I}}}{1+\Sigma e^{a_j+b_j+d_j\bar{I}}}$	$\dfrac{1}{1+\Sigma e^{a_j+b_j+d_j\bar{I}}}$
High	n_H	$\dfrac{e^{a_1+c_1+d_1\bar{I}}}{1+\Sigma e^{a_j+c_j+d_j\bar{I}}}$	$\dfrac{e^{a_2+c_2+d_2\bar{I}}}{1+\Sigma e^{a_j+c_j+d_j\bar{I}}}$	$\dfrac{1}{1+\Sigma e^{a_j+c_j+d_j\bar{I}}}$
Ethnicity				
Indian	n_I	$\dfrac{e^{a_1+d_1+b_1\bar{M}+c_1\bar{H}}}{1+\Sigma e^{a_j+d_j+b_j\bar{M}+c_j\bar{H}}}$	$\dfrac{e^{a_2+d_2+b_2\bar{M}+c_2\bar{H}}}{1+\Sigma e^{a_j+d_j+b_j\bar{M}+c_j\bar{H}}}$	$\dfrac{1}{1+\Sigma e^{a_j+d_j+b_j\bar{M}+c_j\bar{H}}}$
Fijian	n_F	$\dfrac{e^{a_1+b_1\bar{M}+c_1\bar{H}}}{1+\Sigma e^{a_j+b_j\bar{M}+c_j\bar{H}}}$	$\dfrac{e^{a_2+b_2\bar{M}+c_2\bar{H}}}{1+\Sigma e^{a_j+b_j\bar{M}+c_j\bar{H}}}$	$\dfrac{1}{1+\Sigma e^{a_j+b_j\bar{M}+c_j\bar{H}}}$
n	n_T	n_S	n_O	n_N

[a] Summations in the denominators range from $j = 1$ to $j = 2$. Calculation formulae are derived from equations (6.5) in the text. In the last line of the table, n_T denotes total sample size, n_S denotes the sample number of sterilized women, n_O denotes the sample number using some other method, and n_N denotes the sample number using no method.

The results of fitting this model to data from the 1974 Fiji Fertility Survey are shown in Tables 6.2 and 6.3. In Table 6.3 the probabilities are multiplied by 100 and thereby reexpressed as percentages.

Table 6.2 shows the estimated coefficients and their standard errors, which are printed out by the multinomial logit program. The coefficients for equation (6.1a) are given in the column labeled $\log(P_1/P_3)$, and the coefficients for equation (6.1b) are given in the column labeled $\log(P_2/P_3)$. Table 6.3 transforms the results into an MCA format by applying the formulae in Table 6.1 and utilizing the values of the coefficients in Table 6.2 and the mean values of the variables given in the footnote to Table 6.3. Using the formulae in Table 6.1, the coefficients in Table 6.2, and the mean values of the predictor variables given in the footnote to Table 6.3, the reader should recalculate the probabilities in Table 6.3 as a check on understanding.

TABLE 6.2. Multinomial Logit Regression Coefficients for the Model $\log P_j / P_3 = a_j + b_j M + c_j H + d_j I$, $j = 1, 2$: Fecund, Nonpregnant, Currently Married Women Aged 35–44 in the 1974 Fiji Fertility Survey[a]

Predictor Variable	$\log(P_1/P_3)$	$\log(P_2/P_3)$
Intercept	−1.299* (.197)	−1.153* (.197)
Education		
Medium (M)	.373 (.203)	.489* (.216)
High (H)	.169 (.235)	.709* (.234)
Ethnicity		
Indian (I)	1.636* (.192)	.732* (.193)

[a]The underlying model is given in equations (6.1). The three methods are (1) sterilization, (2) other method, and (3) no method. P_1/P_3 compares sterilization with no method, and P_2/P_3 compares other method with no method. Numbers in the $\log(P_1/P_3)$ column are a_1, b_1, c_1, and d_1, and numbers in the $\log(P_2/P_3)$ column are a_2, b_2, c_2, and d_2 [see equations (6.1)]. An asterisk after a coefficient indicates that the coefficient differs from zero with a two-tailed $p < .05$. Numbers in parentheses following coefficients are standard errors. The log likelihood of the test model is $\log L_1 = -982.25$. The log likelihood of the intercept model is $\log L_0 = -1031.20$. The likelihood ratio test of the difference between the test model and the intercept model (see Section 5.3.2 in Chapter 5) yields $\chi^2 = 97.9$ with d.f. = 6, implying that $p < .001$ [see the χ^2 table in Appendix B (Table B.4)]. Pseudo-R = .20.

TABLE 6.3. MCA Table of Adjusted Values of P_j (in Percent) for the Model $\log P_j / P_3 = a_j + b_j M + c_j H + d_j I$, $j = 1, 2$: Fecund, Nonpregnant, Currently Married Women Aged 35–44 in the 1974 Fiji Fertility Survey[a]

Predictor Variable	n	P_1 (Sterilization)	P_2 (Other Method)	P_3 (No Method)
Education				
Low	386	33	22	45
Medium	289	37	28	35
High	179	30	35	35
Ethnicity				
Indian	575	46	24	29
Fijian	379	18	23	59
n	954	335	239	380

[a]Values in this table are obtained using the calculation formulae in Table 6.1, the estimates of coefficients in Table 6.2, and the following mean values of the predictor variables: $\overline{M} = .303$, $\overline{H} = .188$, and $\overline{I} = .603$.

6.1.4. Statistical Inference

Tests of coefficients, tests of the difference between two coefficients, and tests of the difference between two models are done in the same way as in binary logit regression. (See Section 5.3 of Chapter 5.) Some of these tests are illustrated in Table 6.2.

Because the sign of a multinomial logit regression coefficient may not reflect the direction of effect of the predictor variable on either of the probabilities on the left-hand side of the equation, tests of coefficients are usually two-tailed tests.

When testing the difference between two models, one of which is the intercept model, the form of the intercept model, referring back to our earlier example, is

$$\log\frac{P_1}{P_3} = a_1 \tag{6.6a}$$

$$\log\frac{P_2}{P_3} = a_2 \tag{6.6b}$$

$$P_1 + P_2 + P_3 = 1 \tag{6.6c}$$

6.1.5. Goodness of Fit

Pseudo-R^2 is calculated in the same way as in binary logit regression. [See equation (5.56) in Section 5.4 of Chapter 5.] In the multinomial logit example in Tables 6.1–6.3, $k = 6$ and the value of pseudo-R is .20 (see footnote in Table 6.2). The reader can check this result by substituting the log likelihood statistics into equation (5.56).

6.1.6. Changing the Reference Category of the Response Variable

Suppose we ran the model in (6.1) with "other method" instead of "no method" as the reference category. The model would be

$$\log\frac{P_1}{P_2} = e_1 + f_1 M + g_1 H + h_1 I \tag{6.7a}$$

$$\log\frac{P_3}{P_2} = e_2 + f_2 M + g_2 H + h_2 I \tag{6.7b}$$

$$P_1 + P_2 + P_3 = 1 \tag{6.7c}$$

As already mentioned, the estimated values of P_1, P_2, and P_3 for each specified contribution of values of the predictor variables would come out the same as before. However, the coefficients would be different.

It is not necessary to rerun the model with "other method" as the new reference category to estimate and test the coefficients in equations (6.7). Instead, we can derive (6.7) from (6.1). Subtracting equation (6.1b) from (6.1a) and making use of the rule that $\log A - \log B = \log(A/B)$, we obtain

$$\log \frac{P_1}{P_2} = (a_1 - a_2) + (b_1 - b_2)M + (c_1 - c_2)H + (d_1 - d_2)I \tag{6.8a}$$

Comparing (6.7a) and (6.8a), we see that $e_1 = a_1 - a_2$, $f_1 = b_1 - b_2$, $g_1 = c_1 - c_2$, and $h_1 = d_1 - d_2$.

Multiplying both sides of (6.1b) by -1 and making use of the rule that $-\log A = \log(1/A)$, we obtain

$$\log \frac{P_3}{P_2} = -a_2 - b_2 M - c_2 H - d_2 I \tag{6.8b}$$

Comparing (6.7b) and (6.8b), we see that $e_2 = -a_2$, $f_2 = -b_2$, $g_2 = -c_2$, and $h_2 = -d_2$.

In sum, when we change the reference category of the response variable, we can obtain the new coefficients from the original coefficients without rerunning the model.

We can also test the significance of the new coefficients without rerunning the model. In (6.7a) and (6.8a), for example, we wish to test whether the coefficient of M, $f_1 = b_1 - b_2$, differs significantly from zero. In multinomial logit regression as in binary logit regression and ordinary multiple regression, one can optionally print out a regression coefficient covariance matrix, which enables one to compute the standard error of $b_1 - b_2$ as

$$\sqrt{\mathrm{Var}(b_1 - b_2)} = \sqrt{\mathrm{Var}(b_1) + \mathrm{Var}(b_2) - 2\,\mathrm{Cov}(b_1, b_2)}\,.$$

It is then a simple matter to test whether $b_1 - b_2$ differs significantly from zero. The test procedure for the difference between two coefficients is the same as that described earlier in Section 2.9.2 of Chapter 2.

6.2. MULTINOMIAL LOGIT MODELS WITH INTERACTIONS AND NONLINEARITIES

As a more complicated example including interactions and nonlinearities, let us consider the effects of age, age-squared, education (low, medium, high), residence (urban, rural), ethnicity (Indian, Fijian), and residence × ethnicity on contraceptive method choice among fecund, nonpregnant, currently married women aged 15–49. This example is identical to that given in Section 5.5.1 of Chapter 5, except that the response variable is now contraceptive method choice among three methods (sterilization, other method, no method) instead of two methods (use, non-use).

The predictor variables are

A: age
M: 1 if medium education, 0 otherwise
H: 1 if high education, 0 otherwise
U: 1 if urban, 0 otherwise
I: 1 if Indian, 0 otherwise

In log odds form, the model is

$$\log\frac{P_1}{P_3} = a_1 + b_1A + c_1A^2 + d_1M + f_1H + g_1U + h_1I + i_1UI \quad (6.9a)$$

$$\log\frac{P_2}{P_3} = a_2 + b_2A + c_2A^2 + d_2M + f_2H + g_2U + h_2I + i_2UI \quad (6.9b)$$

$$P_1 + P_2 + P_3 = 1 \quad (6.9c)$$

In probability form, the model is

$$P_1 = \frac{e^{a_1+b_1A+c_1A^2+d_1M+f_1H+g_1U+h_1I+i_1UI}}{1 + \sum e^{a_j+b_jA+c_jA^2+d_jM+f_jH+g_jU+h_jI+i_jUI}} \quad (6.10a)$$

$$P_2 = \frac{e^{a_2+b_2A+c_2A^2+d_2M+f_2H+g_2U+h_2I+i_2UI}}{1 + \sum e^{a_j+b_jA+c_jA^2+d_jM+f_jH+g_jU+h_jI+i_jUI}} \quad (6.10b)$$

$$P_3 = \frac{1}{1 + \sum e^{a_j+b_jA+c_jA^2+d_jM+f_jH+g_jU+h_jI+i_jUI}} \quad (6.10c)$$

where the summations range from $j = 1$ to $j = 2$.

TABLE 6.4. Multinomial Logit Regression Coefficients for the Model $\log P_j / P_3 = a_j + b_j A + c_j A^2 + d_j M + f_j H + g_j U + h_j I + i_j UI$, $j = 1, 2$: Fecund, Nonpregnant, Currently Married Women Aged 15–49 in the 1974 Fiji Fertility Survey[a]

Predictor Variable	$\log(P_1/P_3)$	$\log(P_2/P_3)$
Intercept	−18.247* (1.311)	−4.304* (.661)
Age (A)	.856* (.074)	.187* (.042)
Age-squared (A^2)	−.011* (.001)	−.003* (.001)
Medium education (M)	.157 (.127)	.372* (.109)
High education (H)	−.155 (.146)	.552* (.112)
Urban (U)	.869* (.193)	.050 (.144)
Indian (I)	1.740* (.150)	.899* (.110)
Urban × Indian (UI)	−.710* (.228)	.255 (.178)

[a] The underlying model is given in equations (6.9). The three methods are sterilization, other method, and no method. P_1/P_3 compares sterilization with no method, and P_2/P_3 compares other method with no method. An asterisk after a coefficient indicates that the coefficient differs from zero with a two-sided $p < .05$. Numbers in parentheses following coefficients are standard errors. The log likelihood of the test model is $\log L_1 = -3199.44$. The log likelihood of the intercept model is $\log L_0 = -3623.01$. The likelihood ratio test of the difference between the test model and the intercept model is 847.14 with d.f. = 14; consultation of the χ^2 table in Appendix B (Table B.4) indicates $p < .001$. Pseudo-$R = .34$.

The MCA table is constructed by substituting appropriate combinations of ones, zeros, and mean values in equations (6.10). Because A^2 and UI are treated as separate variables, means are calculated as the mean of A^2 (instead of the square of the mean of A) and the mean of UI (instead of the mean of U times the mean of I).

Results of fitting the model, as given in equations (6.9) and (6.10), are shown in Tables 6.4 and 6.5. As a check on understanding, the reader should derive the numbers in Table 6.5, starting from equations (6.10), the fitted coefficients in Table 6.4, and the mean values of the predictor variables in the footnote to Table 6.5.

6.3. A MORE GENERAL FORMULATION OF THE MULTINOMIAL LOGIT MODEL

We now consider the more general case where the response variable has J mutually exclusive and exhaustive categories, denoted $j = 1, 2, \ldots, J$. The Jth category is taken as the reference category for the response variable. Because the ordering of the categories is arbitrary,

TABLE 6.5. MCA Table of Adjusted Values of P_j (in Percent) for the Model $\log P_j / P_3 = a_j + b_j A + c_j A^2 + d_j M + f_j H + g_j U + h_j I + i_j UI$, $j = 1, 2$: Fecund, Nonpregnant, Currently Married Women Aged 15–49 in the 1974 Fiji Fertility Survey[a]

Predictor Variable	n	P_1 (Sterilization)	P_2 (Other Method)	P_3 (No Method)
Age				
15	—	0	23	77
25	—	6	37	57
35	—	31	30	38
45	—	35	23	42
Education				
Low	1266	14	28	58
Medium	1044	14	36	50
High	1169	10	41	48
Residence × ethnicity				
Urban Indian	854	18	44	37
Urban Fijian	375	11	24	65
Rural Indian	1265	18	38	44
Rural Fijian	985	5	25	70
n	3479	715	1095	1669

[a] Values in this table are obtained using equations (6.10), the estimates of coefficients in Table 6.4, and the following means of the predictor variables: $\overline{A} = 31.03$, $\overline{A^2} = 1021.66$, $\overline{M} = .300$, $\overline{H} = .336$, $\overline{U} = .353$, $\overline{I} = .609$, and $\overline{UI} = .245$.

any category can be the Jth category, so that the choice of the reference category is also arbitrary.

In the general case there are also K predictor variables, denoted $X_1, X_2, \ldots, X_K$. The variables X_i may denote not only variables like A, U, and I, but also variables like A^2 and UI.

The multinomial logit model is then specified in log odds form as

$$\log \frac{P_j}{P_J} = \sum_k b_{jk} X_k, \qquad j = 1, 2, \ldots, J - 1 \tag{6.11a}$$

$$\sum_j P_j = 1 \tag{6.11b}$$

where the summation Σ_k ranges from $k = 0$ to $k = K$; where, for convenience, X_0 is defined as $X_0 \equiv 1$; and where the summation Σ_j ranges from $j = 1$ to $j = J$. We define $X_0 \equiv 1$ in order to be able to write the right side of (6.11a) more compactly as a summation of a single term, $b_{jk}X_k$, that includes the intercept as $b_{j0} = b_{j0}X_0$. Equation (6.11a) actually represents $J - 1$ equations. Therefore, equations (6.11a) and (6.11b) together represent J equations, with $(J - 1)(K + 1)$ coefficients to be estimated.

By taking each side of (6.11a) as a power of e and multiplying through by P_J, we can rewrite (6.11a) as

$$P_j = P_J e^{\Sigma_k b_{jk} X_k}, \qquad j = 1, 2, \ldots, J - 1 \tag{6.12a}$$

We also have

$$P_J = P_J \tag{6.12b}$$

We now sum the J equations in (6.12a) and (6.12b) to obtain

$$1 = P_J \sum_j e^{\Sigma_k b_{jk} X_k} + P_J \tag{6.13}$$

where $X_0 = 1$, the summation over j ranges from $j = 1$ to $j = J - 1$, and the summation over k ranges from $k = 0$ to $k = K$. Solving for P_J, we get

$$P_J = \frac{1}{1 + \Sigma_j e^{\Sigma_k b_{jk} X_k}} \tag{6.14}$$

Substituting back into (6.12a) and (6.12b), we obtain the formulation of the model in probability form:

$$\boxed{P_j = \frac{e^{\Sigma_k b_{jk} X_k}}{1 + \Sigma_i e^{\Sigma_k b_{ik} X_k}}, \qquad j = 1, 2, \ldots, J} \tag{6.15}$$

where $X_0 = 1$, the summation Σ_k ranges from $k = 0$ to $k = K$, the summation Σ_i ranges from $i = 1$ to $i = J - 1$, and $b_{J0}, b_{J1}, \ldots, b_{JK}$ are all defined to be zero. This latter definition implies that $e^{\Sigma_k b_{Jk} X_k} = e^0 = 1$, so that (6.15) reduces to (6.14) when $j = J$. The definition $b_{J0} = b_{J1} = \cdots = b_{JK} = 0$ allows the equations for P_j $(j = 1, 2, \ldots, J)$ to be written in the compact one-line form in (6.15).

One interprets and tests this general model in the same way as explained for the simpler examples elaborated earlier in Sections 6.1 and 6.2.

6.4. RECONCEPTUALIZING CONTRACEPTIVE METHOD CHOICE AS A TWO-STEP PROCESS

The multinomial logit models considered above conceptualize contraceptive method choice as a one-step process. A woman perceives three options: sterilization, some other method, or no method. She considers each of the options and then chooses one of them. The act of choosing among the three options is done all in one step.

An alternative conceptualization is that contraceptive method choice is a two-step process: First a woman chooses whether or not to use contraception at all. If the outcome is to use contraception, the second step is to choose a particular method.

The first of these two steps is appropriately modeled by a binary logit model. The second step, if numerous contraceptive methods are considered, is appropriately modeled by a multinomial logit model. In our previous examples from the Fiji survey, however, only two options (sterilization or "other method") remain once the choice to use contraception is made. Thus, in these examples, the second step is also appropriately modeled by a binary logit model.

In the binary logit model of the first of these two steps, let us define the observed value of the response variable to be 1 if the woman chooses to use contraception (without having yet decided which method) and 0 otherwise. Let us denote the predicted value of this response variable by P, the probability of using contraception.

The second binary logit model, pertaining to the second step of choosing a particular method of contraception, is fitted only to the subset of women who actually choose during the first step to use contraception (i.e., to the subset of women whose observed value of the response variable in the first step is 1). In our model of the second step, let us define the observed value of the response variable to be 1 if the woman chooses sterilization and 0 otherwise (otherwise now being "other method"). Let us denote the predicted value of the response variable in this second model by P', the probability of choosing sterilization. P' is a conditional probability. It is conditional on first choosing to use contraception irrespective of method.

Taken together, the results of the two models for the first and second steps imply that the probability that a woman in the original

TABLE 6.6. Combined MCA Results for the Two-Step Choice Model: Fecund, Nonpregnant, Currently Married Women Aged 15–49 in the 1974 Fiji Fertility Survey[a]

Predictor Variable	PP' (Sterilization)	$P(1-P')$ (Other Method)	$1-P$ (No Method)
Age			
15	0	18	82
25	5	40	55
35	32	29	39
45	37	25	38
Education			
Low	16	32	53
Medium	15	39	46
High	10	45	44
Residence × ethnicity			
Urban Indian	19	47	33
Urban Fijian	13	27	60
Rural Indian	20	41	39
Rural Fijian	5	28	67

[a]The only difference between this table and Table 6.5 is that Table 6.5 assumes a one-step process of contraceptive method choice, whereas this table assumes a two-step process. The models used to generate the two tables have the same variables and were applied to the same input data. See text for further explanation.

sample chooses to use contraception and then chooses sterilization is PP', the probability that she chooses to use contraception and then chooses some other method is $P(1 - P')$, and the probability that she chooses not to use contraception at all is $1 - P$. It is easily verified that these three probabilities add to one, as they must.

How do results from the two-step approach compare with our earlier results from the one-step approach? To find out, we redid Table 6.5 using the two-step approach. We first produced two intermediate MCA tables (not shown), one for each of the two binary logit models corresponding to each of the two steps. Thus the first of these two intermediate tables contained estimates of P, and the second contained estimates of P', tabulated in each case by the same predictor variables shown in the row labels of Table 6.5. The probabilities P and P' in these two tables were then combined in the manner described in the previous paragraph. The combined MCA results are shown in Table 6.6, which may be compared with the results in Table 6.5 from the one-step approach. The two tables differ, but not by much. In this example, at least, whether choice is conceptualized as a one-step or a two-step process does not make much difference in the results.

Which model should one use, the one-step model or the two-step model? This question must be answered on the basis of theory and evidence from previous studies. In the case of contraceptive use in Fiji in 1974, however, the choice between the two models is not clear. It seems plausible that many women decided in two steps, in which case a two-step model seems justified. But for others the decision whether to use contraception at all may have been affected by what methods were locally available, in which case a one-step model seems justified. Without more evidence, we cannot say which mode of choice behavior predominates. In the absence of such evidence, it seems prudent to estimate both models and then compare the results.

In a country like the United States, where a wide variety of contraceptive methods are universally available, the decision whether to use contraception at all is not affected by availability of methods, so that a two-step model is more clearly appropriate.

6.5. FURTHER READING

More advanced treatments may be found in Theil (1969), Cragg and Uhler (1970), Schmidt and Strauss (1975), Amemiya (1981), and Maddala (1983).

7

SURVIVAL MODELS, PART 1: LIFE TABLES

Survival models are applied to data that specify the time elapsed until an event occurs. The concept of "time elapsed" implies a starting event and a terminating event. Examples are time elapsed between birth and death, time elapsed between divorce and remarriage, or time elapsed between 15th birthday and first full-time job.

A *cohort* is a set of individuals, defined by some set of characteristics, who have experienced the starting event in a specified time period. Terminating events are generically called *failures*, and the times (measured from the starting event) at which the failures occur are called *failure times*. A *life table* is a statistical presentation of the life history of a cohort, commencing with the starting event, as the cohort is progressively thinned out over time by failures.

The life table is a basic building block for hazard models, which are considered in Chapters 8 and 9. This chapter, which is preparatory, is limited to those aspects of the life table that are needed for hazard models.

We consider two ways of calculating the life table: (1) the actuarial method and (2) the product-limit method. The product-limit method is the method that is actually used in the estimation of hazard models. We additionally present the actuarial method by way of introduction to life tables, because the actuarial method is easier to understand than the product-limit method.

7.1. ACTUARIAL LIFE TABLE

As a simple example, let us consider a life table of transitions from fifth to sixth birth, derived from reproductive histories of women in the 1974 Fiji Fertility Survey. The starting event is a woman's fifth birth, and the terminating event is her sixth birth. The cohort under consideration is all women in the Fiji survey who ever had a fifth birth. Recall from Chapter 3 that mean completed family size in the Fiji sample was about six children per woman. We therefore expect fairly high rates of transition from fifth to sixth birth.

The first step in constructing an actuarial life table is to define time intervals. Because the survey reports date of birth to the nearest month, it is convenient to define one-month time intervals, with time measured from the fifth birth. We shall terminate the life table at 10 years, or 120 months. If a sixth birth does not occur within 120 months after the fifth birth, it is unlikely to occur at all. Thus the choice of a 120-month cutoff is not completely arbitrary.

The second step is to define the life-table quantities to be estimated. In the following definitions, "starting event" means fifth birth, "failure" means sixth birth, and "survival" means not yet having had a sixth birth.

- t: time elapsed since the starting event, in months
- q_t: estimated conditional probability of failure between t and $t + 1$, given survival to t (the interval t to $t + 1$ ranges from t up to but not including $t + 1$)
- p_t: estimated conditional probability of surviving from t to $t + 1$, given survival to t
- S_t: estimated unconditional probability of survival from the starting event to time t
- F_t: estimated unconditional probability of failure between the starting event and time t

Additional life-table quantities can be defined, but we shall not consider them in this book.

The interrelationships between q_t, p_t, S_t, and F_t, which we shall discuss next, are illustrated in Table 7.1, which shows in detail the first 24 months of the actuarial life table for indigenous Fijians in Fiji. Later we shall present in considerably less detail life tables out to 120 months for both indigenous Fijians and Indians.

TABLE 7.1. First 24 Months of Actuarial Life Table of Transitions from Fifth to Sixth Birth: Indigenous Fijians, 1974 Fiji Fertility Survey

t	q_t	p_t	S_t	F_t	λ_t
0	0	1	1	0	0
1	0	1	1	0	0
2	0	1	1	0	0
3	0	1	1	0	0
4	0	1	1	0	0
5	0	1	1	0	0
6	0	1	1	0	0
7	0	1	1	0	0
8	.0059	.9941	1	0	.0059
9	.0030	.9970	.9941	.0059	.0030
10	.0045	.9955	.9912	.0087	.0045
11	.0122	.9878	.9867	.0133	.0123
12	.0124	.9876	.9747	.0253	.0125
13	.0143	.9857	.9626	.0374	.0144
14	.0097	.9903	.9488	.0512	.0098
15	.0181	.9819	.9396	.0604	.0183
16	.0084	.9916	.9226	.0774	.0085
17	.0291	.9709	.9148	.0852	.0295
18	.0159	.9841	.8882	.1118	.0160
19	.0270	.9730	.8741	.1259	.0273
20	.0315	.9685	.8505	.1495	.0320
21	.0365	.9635	.8237	.1763	.0372
22	.0322	.9678	.7936	.2064	.0327
23	.0313	.9687	.7681	.2319	.0318
24	.0434	.9566	.7440	.2560	.0444

From the definitions of q_t, p_t, S_t, and F_t, it is immediately clear that

$$\boxed{p_t + q_t = 1} \tag{7.1}$$

$$\boxed{S_t + F_t = 1} \tag{7.2}$$

Let us suppose that the probabilities q_t are given. From the probabilities q_t, one can calculate p_t, S_t, and F_t as follows: First, from equation (7.1), it is evident that the probabilities p_t can be calculated from the probabilities q_t as

$$\boxed{p_t = 1 - q_t} \tag{7.3}$$

The probability of surviving to time 0 is 1, by definition. Thus

$$\boxed{S_0 = 1} \tag{7.4}$$

The probability of surviving to the start of each subsequent month can be calculated iteratively as

$$\begin{aligned} S_1 &= S_0 p_0 \\ S_2 &= S_1 p_1 \\ &\vdots \\ S_{120} &= S_{119} p_{119} \end{aligned} \tag{7.5}$$

In each case the unconditional probability of surviving to the start of a month is calculated as the product of the unconditional probability of surviving to the start of the previous month and the conditional probability of surviving from the start of the previous month to the start of the current month. We can write (7.5) more concisely as

$$\boxed{S_{t+1} = S_t p_t} \tag{7.6}$$

By successive substitution in equations (7.4) and (7.5), we can write (7.6) alternatively as

$$\boxed{S_{t+1} = \prod_{i=0}^{t} p_i} \tag{7.7}$$

where $\prod$ is the product symbol; thus $S_1 = p_0$, $S_2 = p_0 p_1$, $S_3 = p_0 p_1 p_2$, and so on.

Once the probabilities S_t are known, then, from (7.2), the probabilities F_t can be calculated as

$$\boxed{F_t = 1 - S_t} \tag{7.8}$$

Knowing the probabilities q_t enables us to calculate the rest of the life table, but how do we link the probabilities q_t to the data? The data in this case are birth histories for each ever-married woman in the Fiji Fertility Survey. The birth history for a woman specifies the date of each of her births that occurred before the survey. For a woman who had a fifth birth, it is a simple matter to calculate from these dates the value of t for her sixth birth (if there was one), as measured from the time of her fifth birth.

First we define the following quantities pertaining directly to the birth history data:

t: time since the birth of her fifth child

N_0: number of women in the survey who had at least five births

N_t: number of survivors at time t (women with five births who have not yet had a sixth birth)

D_t: number of failures during the interval t to $t + 1$ (failures—i.e., sixth births—are analogous to deaths, hence the notation D)

c_t: number of women who were *censored* during the interval t to $t + 1$ ("censored" means "lost to observation," due to reaching interview before the end of the interval and before having a sixth birth[1])

If there were no censoring (i.e., if all women in the sample who had a fifth birth also had a sixth birth before the survey), the link between q_t and the data would be simple:

$$q_t = \frac{D_t}{N_t} \text{ in the absence of censoring} \tag{7.9}$$

The link is somewhat more complicated in the presence of censoring. We assume that, on the average, the c_t women who were censored in the interval t to $t + 1$ reached interview at the midpoint of the interval, implying an observed exposure to the risk of failure amounting to one-half of the interval. *Equivalently, we can imagine that* $.5c_t$ *women were exposed over the entire interval, and that the other* $.5c_t$ *women were not exposed at all during the interval. Thus the effective (fully exposed) N at the start of the interval is* $N_t' = N_t - .5c_t$. Thus, in

[1]Regarding censoring, note that the Fiji data are *retrospective data*. In *prospective studies*, where the sample members are followed over time, censoring can occur not only because of survival to the end of the data collection process or to the start of analysis, but also because of death or loss to follow-up.

the presence of censoring, the link between q_t and the data is

$$q_t = \frac{D_t}{N'_t} \text{ in the presence of censoring} \tag{7.10}$$

Another important quantity is the *hazard* (also called the *hazard rate*), denoted λ_t. *The hazard is a time-specific failure rate, defined as failures per woman per month.* The hazard λ_t is calculated as the observed number of failures during the interval divided by the product of the time width of the interval (in this case, one month) and the average number alive during the interval, adjusted for censoring:

$$\lambda_t = \frac{D_t}{.5[N'_t + (N'_t - D_t)]} \tag{7.11}$$

After dividing numerator and denominator by N'_t and substituting q_t for D_t/N'_t and p_t for $1 - q_t$, (7.11) becomes

$$\lambda_t = \frac{2q_t}{1 + p_t} \tag{7.12}$$

The set of hazard rates over all the time intervals is called the *hazard function*.

By substituting $1 - q_t$ for p_t, one can also solve (7.12) for q_t in terms of λ_t:

$$q_t = \frac{2\lambda_t}{2 + \lambda_t} \tag{7.13}$$

If hazard rates have already been calculated from the data using the formula in (7.11), then (7.13) provides an alternative way of linking q_t to the data.

Note that, if any one of the quantities q_t, p_t, S_t, F_t, or λ_t is given for each t, it is a simple matter to calculate all the remaining life-table quantities. For example, if S_t is known, one can calculate $F_t = 1 - S_t$, $p_t = S_{t+1}/S_t$, $q_t = 1 - p_t$, and $\lambda_t = 2q_t/(1 + p_t)$.

We often wish to summarize the life table with a single number. In this example, one might use F_{120}, which is the estimated probability of

TABLE 7.2. Among Women Who Have Had Five or More Children, the Estimated Life-table Probability of Not Yet Having Had a Sixth Child, S_t, by Ethnicity and Life-table Calculation Method: 1974 Fiji Fertility Survey[a]

Months Since Fifth Birth	Actuarial Method		Product-Limit Method	
	Indians	Fijians	Indians	Fijians
0	1.0000	1.0000	1.0000	1.0000
12	.9374	.9747	.9154	.9627
24	.6089	.7440	.5753	.7124
36	.3786	.3855	.3712	.3568
48	.3047	.2491	.3016	.2444
60	.2684	.1954	.2693	.1923
72	.2357	.1644	.2318	.1606
84	.2110	.1417	.2119	.1428
96	.1967	.1355	.1977	.1366
108	.1799	.1315	.1811	.1326
120	.1766	.1315	.1778	.1326

[a]The life tables were calculated by month. Selected values of S_t for every 12th month are shown.

progressing from fifth to sixth birth within 120 months after the fifth birth.

Why not calculate the progression ratio F_{120} directly, without the life table? We could do this, but we would have to restrict the calculation to those women who were under observation for a full 10 years following the fifth birth. For this smaller set of women, we could calculate the fraction that had a sixth birth during the 10-year period following the fifth birth. *In this direct calculation, data on censored women (women interviewed and therefore lost to observation before* 10 *years elapsed) are thrown out and therefore wasted. In the life table calculation, on the other hand, data on the censored women are fully utilized.* Another important advantage of the life table is that it provides information on the time distribution of sixth births within the 10-year period following the fifth birth.

The left half of Table 7.2 presents selected values of the S_t column up to 120 months, calculated by the actuarial method life table for Indians and indigenous Fijians separately.

7.1.1. Statistical Inference

Under the assumption that the underlying population is *homogeneous* (meaning that each woman in the population has the same set of underlying time-specific risks of having a sixth birth—i.e., the same hazard function), approximate standard errors for S_t and λ_t can be calculated and printed out by the computer program (in this case, BMDP). As usual in this book, we shall not be concerned about how these standard errors are derived, but instead shall focus on how to use and interpret them.

For large samples, the sampling distributions of S_t and λ_t are approximately normal, so that confidence intervals can be constructed and statistical tests conducted in the usual way.

Suppose, for example, that we have calculated two life tables of transitions from fifth to sixth birth, one for Indians and one for indigenous Fijians, and we wish to test whether F_{120} for Indian differs significantly from F_{120} for Fijians. This is equivalent to a test of whether S_{120} for Indians (denoted "Ind") differs significantly from S_{120} for Fijians (denoted "Fiji"), because $F_{120}(\text{Ind}) - F_{120}(\text{Fij}) = [1 - S_{120}(\text{Ind})] - [1 - S_{120}(\text{Fij})] = S_{120}(\text{Fij}) - S_{120}(\text{Ind})$.

Because Indians and Fijians are sampled independently, $\text{Cov}[S_{120}(\text{Fij}), S_{120}(\text{Ind})] = 0$ over repeated samples from the same underlying population. Therefore,

$$\begin{aligned}\text{Var}[F_{120}(\text{Ind}) - F_{120}(\text{Fij})] &= \text{Var}[S_{120}(\text{Fij}) - S_{120}(\text{Ind})] \\ &= \text{Var}[S_{120}(\text{Fij})] + \text{Var}[S_{120}(\text{Ind})] \quad (7.14)\end{aligned}$$

$\text{Var}[S_{120}(\text{Fij})]$ is the square of the standard error of S_{120} for Fijians, and $\text{Var}[S_{120}(\text{Ind})]$ is the square of the standard error of S_{120} for Indians. These standard errors are printed out along with the life tables. One can therefore construct a Z value as

$$Z = \frac{F_{120}(\text{Ind}) - F_{120}(\text{Fij})}{\sqrt{\text{Var}[S_{120}(\text{Ind})] + \text{Var}[S_{120}(\text{Fij})]}} \quad (7.15)$$

This Z value, which for large samples has an approximately normal sampling distribution, can be used in the usual way to test whether $F_{120}(\text{Ind}) - F_{120}(\text{Fij})$ differs significantly from zero, which is equivalent to testing whether $F_{120}(\text{Ind})$ differs significantly from $F_{120}(\text{Fij})$.

When life tables for Indians and Fijians are calculated from the Fiji data, the following values, among others, are printed out: $F_{120}(\text{Ind}) = .8234$, $\text{Var}[S_{120}(\text{Ind})] = (.0148)^2$, $F_{120}(\text{Fij}) = .8685$, and $\text{Var}[S_{120}(\text{Fij})] = (.0156)^2$. Equation (7.15) then becomes

$$Z = \frac{.8234 - .8685}{\sqrt{(.0148)^2 + (.0156)^2}} = -2.10 \quad (7.16)$$

Consulting the table of normal probabilities in Appendix B (Table B.2), we find that $Z = -2.10$ corresponds to a two-tailed $p = .04$. We conclude that the difference between Indians and Fijians in F_{120} (the probability of having a sixth birth within 120 months of a fifth birth) is significant at the 5 percent level, because $p < .05$.

Statistical tests are not usually performed on the hazard rates, λ_t, because each hazard rate pertains to a small time interval and is therefore based on a comparatively small number of failures. A cumulative measure such as F_{120} is based on many more failures and is therefore more statistically stable.

7.2. PRODUCT-LIMIT LIFE TABLE

The product-limit life table is also referred to as the *Kaplan–Meier* life table, in recognition of the inventors of the product-limit approach.

To illustrate, let us recalculate our Fiji life table using the product-limit method. Values of the survival function S are now calculated at each unique failure time. Time intervals are therefore irregular and determined by the data.

It is convenient to distinguish between survival times and failure times. *Survival times* include both failure times and censoring times. Let n denote the number of survival times, which is the same as the size of the sample, because every sample observation has either a failure time or a censoring time. The survival times may be ordered from smallest to largest: $t_1 \leq t_2 \leq \cdots \leq t_n$. If two or more survival times are tied, then, for reasons to be clarified shortly, the ordering of observations within the tie is arbitrary, except that tied failure times are placed ahead of tied censoring times within the overall tie.

The subscript of S_t now denotes time elapsed since the starting event. By definition, $S_0 \equiv 1$. Subsequently, the estimate of S is calculated and printed out at each unique failure time:

$$\boxed{S_t = \prod_{t_i < t} \frac{n - i + 1 - \delta_i}{n - i + 1}} \tag{7.17}$$

where n denotes the number of survival times (same as sample size), t_i denotes the ith survival time, and δ_i is 1 if t_i represents a failure and 0 if t_i represents a censored observation. If a tie includes failures as well as censored observations, then, as already mentioned, the failures are assumed to occur immediately before the censored observations. That way, the censored individuals are correctly included in the population at risk of failure at the start of the previous interval when applying (7.17).

As a hypothetical example, suppose that $n = 500$; t_1 and t_2 represent deaths; t_3 represent a censored observation; and t_4 and t_5 represent tied deaths so that t_4 and t_5 coincide. Applying (7.17), we obtain

$$S_0 = 1 \tag{7.18}$$

$$S_{t_1} = 1 \tag{7.19}$$

$$S_{t_2} = 499/500 = .9980 \tag{7.20}$$

$$\left[S_{t_3} = (499/500)(498/499) = .9960\right] \tag{7.21}$$

$$S_{t_4} = (499/500)(498/499)(498/498) = .9960 \tag{7.22}$$

$$\begin{aligned} S_{t_6} &= (499/500)(498/499)(498/498)(496/497)(495/496) \\ &= .9920 \end{aligned} \tag{7.23}$$

The entry for S_{t3} is shown in square brackets in (7.21) because t_3 is not a failure time. There is no entry for S_{t5} in the above calculations because t_4 and t_5 are tied. The graph of the results in (7.18)–(7.23) is a step function, as shown in Figure 7.1. In the graph, a death at t_i produces a drop to the next lower step immediately after t_i. (By "immediately" is meant an infinitesimally small time interval.)

In the case of our Fiji example, survival times are reported to the nearest month, so that the data contain many ties. Selected values of the survival function from the product-limit life table are shown in the right half of Table 7.2 for Indians and Fijians separately. Standard errors for S_t are also printed out by most computer programs, although they are not shown in Table 7.2.

The significance of the product-limit life table lies in the fact that it provides the basis for estimating hazard models, which may be viewed as multivariate life tables. The product-limit formulation of the life table facilitates the formulation of a likelihood function for the hazard model, so that the model can be estimated and its statistical properties ascertained.

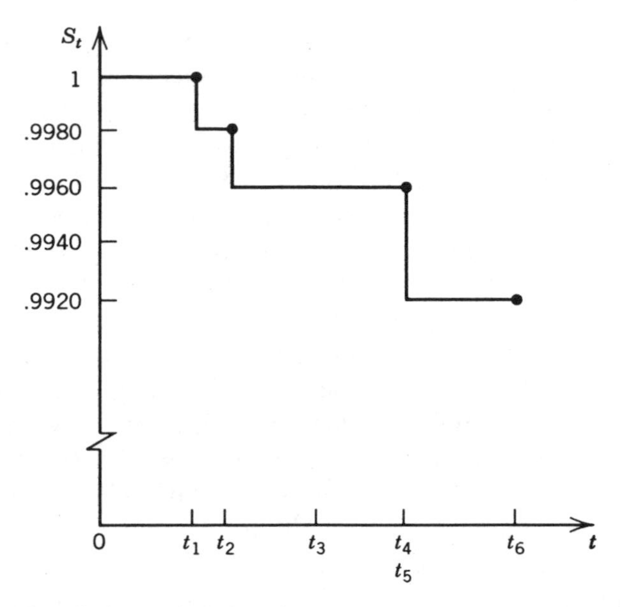

Figure 7.1. Product-limit survival function, S_t, as given by equations (7.18)–(7.23). **Notes**: The times t_i are in general irregularly spaced. The dots in the step function indicate that the function is continuous from the left at the points of discontinuity.

7.2.1. Testing the Difference Between Two Survival Curves

Two tests are commonly used, the Mantel test and the Breslow test.

The *Mantel test* (also called the *Mantel–Cox test* or the *generalized Savage test*) was suggested by Mantel (1966), and an alternative derivation of it was proposed by Cox (1972). The test is based on a series of χ^2 statistics for 2×2 tables constructed for failure time intervals. Suppose, for example, that we wish to compare the survival curves for Indians and Fijians. If survival times are rounded to the nearest month, a 2×2 table can be constructed for each consecutive month. For each such table, column labels can be chosen as Indians and Fijians, and row labels can be chosen as surviving and dead. In the Mantel method, the χ^2 statistics are pooled to yield a single statistic that can be used for testing. In practice, the user need only understand the general idea of the test, because most computer programs compute the statistics, including a p value. If $p < .05$, the two survival curves for Indians and Fijians differ significantly from each other at the 5 percent level.

The *Breslow test* (also called the *generalized Wilcoxon test* or the *generalized Gehan test*) is based on an ordering of the failure times, from earlier to later. To continue with the example of Indians and

Fijians, one first constructs an ordering of individual failure times. The ordering is a single ordering that includes both Indians and Fijians. From this ordering, the Breslow test checks to see if the failure times of individuals in one group, say Indians, tend to occur earlier or later than the failure times of the other group, Fijians. This method has been generalized by Gehan (1965) to allow for censoring, and by Breslow (1970) to permit comparison of more than two groups. Again, the user need only understand the general idea of the test, because most computer programs do the computations.

These tests may give quite different results when the survival curves cross over each other, as is the case with the Indian and Fijian survival curves in Table 7.2. (Indians start out with lower survival but end up with higher survival than Fijians.) In this case the Mantel test yields $p = .919$, and the Breslow test yields $p = .002$. The Breslow test does a better job of detecting the large difference between the two survival curves. However, crossovers are unusual. The more usual case is that hazard rates are consistently higher for one of the two groups being compared. Life tables for urban and rural in Fiji are an example. (Survival curves for urban and rural are shown later in Figure 8.2 in Chapter 8.) The Mantel test yields $p = .0002$, and the Breslow test yields $p = .0012$. In this case the results of the two tests are similar.

7.3. THE LIFE TABLE IN CONTINUOUS FORM

In Chapters 8 and 9, we shall find it useful to represent the life table in continuous form (i.e., in the form of smooth curves), even though actual calculations are based on discrete data. In this chapter, the discussion of the continuous form of the life table is limited to derivation of $S(t)$ from $\lambda(t)$, where $S(t)$ is the survivorship function and $\lambda(t)$ is the instantaneous hazard rate at time t. This discussion also introduces the concept of the cumulative hazard.

The continuous-form derivation of $S(t)$ from $\lambda(t)$ requires calculus. For an infinitesimally small time interval centered on t, $\lambda(t)$ may be written as

$$\lambda(t) = \frac{-dS}{S\,dt} = -\frac{1}{S}\frac{dS}{dt} \tag{7.24}$$

The quantities dt and dS are called *differentials*. Differentials are infinitesimally small. The differential dt is the width of the infinitesimally small time interval, and dS is the change in S as t increases by the increment dt over the interval. Because S decreases as t increases, dS is negative. Therefore $-dS$ is positive. $-dS$ represents failures in the interval. Therefore, $-dS/(S\,dt)$ has the form of a hazard rate (failures per person per unit of time). $-dS/(S\,dt)$ can be rewritten as $-(1/S)(dS/dt)$, where dS/dt is the derivative of S with respect to t.

If we multiply through by $-dt$, (7.24) becomes

$$\frac{dS}{S} = -\lambda(t)\,dt \tag{7.25}$$

Integrating from 0 to t, and recalling that $S(t) = 1$ at $t = 0$ and that $\log(1) = 0$, we get

$$\log S(t) = -\int_0^t \lambda(x)\,dx \tag{7.26}$$

where x is chosen as the variable of integration in order to distinguish it from t in the limits of integration. Taking both sides of (7.26) as a power of e, we obtain

$$\boxed{S(t) = e^{-\int_0^t \lambda(x)\,dx}} \tag{7.27}$$

Let us now define the *cumulative hazard*, $\Lambda(t)$, as

$$\boxed{\Lambda(t) \equiv \int_0^t \lambda(x)\,dx} \tag{7.28}$$

(Λ is uppercase lambda.) Then (7.27) can also be written

$$\boxed{S(t) = e^{-\Lambda(t)}} \tag{7.29}$$

Solving this equation for $\Lambda(t)$, we obtain another useful formula,

$$\Lambda(t) = -\log S(t) \tag{7.30}$$

We can express the cumulative hazard in discrete form as

$$\Lambda(t) = \int_0^t \lambda(x)\,dx \approx \sum_{x<t} \lambda_x \,\Delta x \tag{7.31}$$

where λ_t is the discrete time-specific failure rate for the time interval t to $t + \Delta t$. If we sum over one-month time intervals, $\Delta t = 1$, so that (7.31) can be rewritten

$$\boxed{\Lambda(t) \approx \sum_{x=0}^{t-1} \lambda_x} \tag{7.32}$$

where we now interpret λ_t as the hazard rate for the interval t to $t + 1$. Note that $\Lambda(0) = 0$.

7.4. FURTHER READING

For simple explanations of the life table that go beyond this chapter, see Barclay (1958) and Benedetti, Yuen, and Young (1981). For more advanced treatments, see Chiang (1968), Kalbfleisch and Prentice (1980), and Cox and Oakes (1984). For more detailed explanations of the Mantel test, see Mantel (1966) and Cox (1972). For more detailed explanations of the Breslow test, see Breslow (1970) and Gehan (1965).

8

SURVIVAL MODELS, PART 2: PROPORTIONAL HAZARD MODELS

A *proportional hazard model* (also called a *proportional hazards model*) may be viewed as a multivariate life table. In conventional life-table analysis, if we want to know, say, the proportion surviving to some specified time, cross-classified by residence and education, we must divide the sample into cross-tabulated subsamples and calculate a separate life table for each subsample. Frequently we can't do this, because we don't have enough cases.

In a proportional hazard model, we fit a multivariate life table to the entire sample. The model is a simplified version of reality, whereby the hazard function (from which the rest of the life table can be calculated) is assumed to depend on residence and education according to a simple mathematical rule. If the model fits the data, then, in effect, the multivariate hazard model economizes on cases, so we don't need such a large sample to get statistically meaningful results. Note, by analogy, that this is the same advantage that multiple regression enjoys over simple cross-tabulation.

In ordinary life-table analysis, we assume that the population is *homogeneous*. By "homogeneous" we mean that each sample member has the same underlying hazard function. In proportional hazard models, on the other hand, the hazard is not only a function of time but also a function of specified predictor variables, such as residence and education. Thus the proportional hazard model allows for a certain amount of *heterogeneity*. There will still be *unobserved hetero-*

geneity, because only a limited number of individual characteristics can be measured and included in the model.

Although the proportional hazard model may be viewed as a multivariate life table, its main use is not to calculate life tables but rather to assess the effects of the predictor variables on the hazard function. *The hazard function (consisting of the entire set of time-specific hazard rates) is viewed as the response variable, and variables such as residence and education are the predictor variables.*

The proportional hazard model is based on the product-limit life table and handles censoring in the same way as the product-limit life table. Input data specify the following: time elapsed between the starting event and the terminating event or censoring; status at last observation (failure or censored); and values of the predictor variables for each individual. The proportional hazard model is fitted by the method of maximum likelihood.

8.1. BASIC FORM OF THE PROPORTIONAL HAZARD MODEL

It is convenient to develop the proportional hazard model using continuous functions, even though, for practical reasons, the model is ultimately fitted to discrete data. This means, for example, that we conceptualize the survival function as a continuous curve, $S(t)$, instead of a step function, S_t. We similarly conceptualize the hazard as a continuous function, $\lambda(t)$, which may be thought of as a time-specific hazard rate (failures per person per unit time) for an infinitesimally small time interval centered on time t. Alternatively, at the individual level, $\lambda(t)$ may be thought of as the probability of failure per unit time. [It then follows that the probability of failure during a small time interval, say from $t - .1$ to $t + .1$, is approximately $.2\lambda(t)$—the width of the interval times $\lambda(t)$.]

The general form of the hazard model is

$$\lambda_x(t) = \lambda_0(t)C_x(t) \tag{8.1}$$

where $\lambda_0(t)$ denotes a *baseline hazard function*, x denotes a set of characteristics (such as urban residence and low education), and $C_x(t)$ is a multiplier specific to persons with the set x of characteristics.

If the model can be written

$$\lambda_x(t) = \lambda_0(t)C_x \tag{8.2}$$

so that the multiplier C_x is constant over time, then the model is called a *proportional hazard model*, with $\lambda_x(t)$ proportional to $\lambda_0(t)$ and C_x the constant of proportionality. For example, if $C_x = 1.7$, then, regardless of the amount of time elapsed since the starting event, the hazard rate for those with the set x of characteristics is 1.7 times the baseline hazard rate. [Recall the definition of proportionality: Two variables X and Y are proportional if the proportion Y/X is constant for all values of (X, Y); if the constant is denoted by k, then $Y/X = k$, or, equivalently, $Y = kX$, where k is the *constant of proportionality*.]

8.1.1. A Simple Example

Suppose a proportional hazard model of transitions from fifth to sixth birth, as in the previous chapter, and let the population characteristic be urban–rural residence. In equation (8.2), we can think of x as taking on two possible values, urban or rural. C_x then takes on two possible values, C_U or C_R. For convenience, let us choose our baseline hazard function $\lambda_0(t)$ as the hazard function for rural. Then $C_R = 1$. Now we need only concern ourselves with estimating C_U.

Clearly C_U must be a positive number. As shown in Figure 8.1, any positive number can be represented as e^Z for some Z, because e^Z ranges from (but not including) zero to infinity (∞) as Z ranges from $-\infty$ to $+\infty$. Suppose that C_U is a number less than 1 (because urban fertility is usually lower than rural fertility), and let us denote the

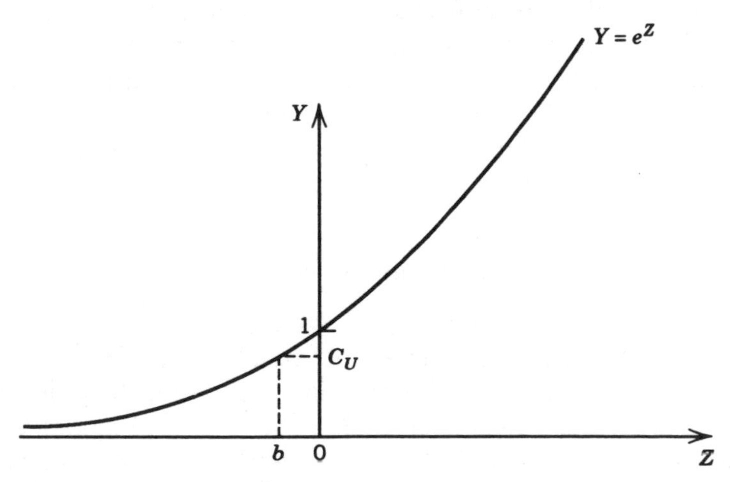

Figure 8.1. Values of e^b and b corresponding to the multiplier C_U.

desired value of Z corresponding to C_U by b. We can obtain b from the graph in Figure 8.1 by locating the value of e^Z that equals C_U and then determining b as the corresponding value of Z. Because, in this example, $e^b < 1$, it follows that $b < 0$. Algebraically, $e^b = C_U$ and therefore $b = \log C_U$.

The model now consists of two equations:

$$\lambda_r(t) = \lambda_0(t) \tag{8.3}$$

$$\lambda_u(t) = \lambda_0(t)e^b \tag{8.4}$$

By employing a dummy variable U (1 if urban and 0 otherwise), we can represent (8.3) and (8.4) more succinctly as

$$\lambda(t) = \lambda_0(t)e^{bU} \tag{8.5}$$

If $U = 0$, corresponding to rural, (8.5) reduces to (8.3). If $U = 1$, corresponding to urban, (8.5) reduces to (8.4).

Why bother to put the model in the form of (8.5)? Why not just stay with (8.2), which seems simpler? A very important reason for using (8.5) is that it allows specification of the residence variable in the form to which we are accustomed, namely a dummy variable U. This is convenient for both estimation and analysis. The use of the exponential term, which is always positive, ensures that the ratio $\lambda(t)/\lambda_0(t) = e^{bU}$ will always be positive, as it should be.

But why e^{bU}? Why not 2^{bU} or 7^{bU}? Why use e as the base of the exponential term when any other positive number would appear to serve equally well? The reason is that the number e has properties that simplify the mathematics involved in estimating the hazard model from sample data. Thus the choice of e is a matter of convenience.

Note that the coefficient of U in equation (8.5) is denoted by the Roman letter b, indicating a sample estimate. Throughout this chapter and the next, hazard models are stated in estimated form.

8.1.2. Addition of a Second Predictor Variable

Let us now expand the model by adding an additional predictor variable, E, defined as the number of completed years of education.

In Section 8.1.1, we saw that a one-unit increase in U multiplies the hazard by e^b. By analogy, we can define the expanded model so that a one-year increase in E multiplies the hazard by e^c for some value of c. To keep the model as simple as possible, we assume that each additional year of education has the same multiplicative effect on the hazard. (Note the parallel to ordinary multiple regression, where we assume that each additional unit increase in a predictor variable produces the same additive effect on the response variable.) It follows that the effect of two years of education is to multiply the hazard by $e^c e^c = e^{2c}$. In general, the effect of E years of education is to multiply the hazard by e^{cE}.

As a further simplification, let us assume that the effect of E on the hazard does not depend on the level of U (i.e., there is no interaction between E and U). We can then write the expanded model as

$$\lambda(t) = \lambda_0(t)e^{bU}e^{cE}$$
$$= \lambda_0(t)e^{bU+cE} \tag{8.6}$$

In (8.6), the baseline hazard is interpreted as the hazard for rural persons with no education, obtained by setting both $U = 0$ and $E = 0$ in (8.6). Equation (8.6) then reduces to $\lambda(t) = \lambda_0(t)$. *The baseline hazard, $\lambda_0(t)$, is interpreted as the estimated hazard for persons for whom all the predictor variables are zero*. In this regard, the baseline hazard, $\lambda_0(t)$, is analogous to the intercept term in multiple regression.

8.1.3. Redefinition of the Baseline Hazard

Defining the baseline hazard as the hazard for persons for whom all the predictor variables are zero can lead to conceptually awkward (though still valid) results. Suppose, for example, that we wish to estimate the model in (8.6) for women in the United States who have had at least five births. The difficulty is that virtually all women in the United States have had at least one year of schooling. Therefore the baseline hazard $\lambda_0(t)$ refers to a group of women (rural women with no education) that for all practical purposes does not exist. In this instance $\lambda_0(t)$ must be viewed as an extrapolation. (An analogous situation arises in bivariate linear regression when we extrapolate the regression line $Y = a + bX$ and use it to calculate a value of Y for a value of X that is outside the range of the data.)

To get around the problem of a conceptually awkward baseline hazard, we employ a trick to redefine the baseline hazard. Employing the fact that $e^{bU'+cE'}e^{-(bU'+cE')} = 1$, we rewrite (8.6) as

$$\lambda(t) = \lambda_0(t)e^{bU'+cE'}e^{-(bU'+cE')}e^{bU+cE} \tag{8.7}$$

where U' and E' are arbitrarily chosen values of U and E. Next we define

$$\boxed{\lambda'(t) \equiv \lambda_0(t)e^{bU'+cE'}} \tag{8.8}$$

$$\boxed{a \equiv -(bU' + cE')} \tag{8.9}$$

Equation (8.7) can then be written

$$\lambda(t) = \lambda'(t)e^{a}e^{bU+cE} \tag{8.10}$$

or

$$\boxed{\lambda(t) = \lambda'(t)e^{a+bU+cE}} \tag{8.11}$$

where $\lambda'(t)$ *is a redefined baseline hazard function pertaining to persons for whom* $U = U'$ *and* $E = E'$. *The quantities* U' *and* E' *are called the baseline values of* U *and* E. For example, we could define the baseline hazard as the hazard for urban persons with 5 years of education, so that $U' = 1$ and $E' = 5$. *It is clear from equations* (8.7)–(8.11) *that the choice of a baseline hazard is arbitrary, insofar as the choice of baseline values of* U *and* E *is arbitrary*. (The baseline hazard is by no means completely arbitrary, however, because it still must be fitted to the data, as discussed later.) If we set $U' = E' = 0$, then $a = 0$, $\lambda'(t) = \lambda_0(t)$, and (8.11) reduces to (8.6).

The derivation of (8.11) *from* (8.6) *indicates that changing the specification of the baseline hazard function* (*i.e.*, *changing the values of* U' *and* E') *has no effect on the fitted values of* b *and* c. Equation (8.9) shows, however, that *the value of the constant term* a *depends on how the baseline hazard is defined*.

Because a is somewhat of a nuisance term, it is useful to reexpress (8.11) in another form that does not explicitly include the constant term. Substituting $a = -(bU' + cE')$ into (8.11), we obtain

$$\lambda(t) = \lambda'(t)e^{-(bU'+cE')+bU+cE} \tag{8.12}$$

With some rearranging of terms, this becomes

$$\boxed{\lambda(t) = \lambda'(t)e^{b(U-U')+c(E-E')}} \tag{8.13}$$

Equation (8.13) expresses the hazard model in *deviation form*.

We choose the baseline values U' and E' on the basis of analytical convenience. *In most analyses, we have no preference for particular baseline values of U and E, so we set them to their mean values,* $\overline{U}$ *and* $\overline{E}$*, so that* $\lambda'(t)$ *becomes* $\overline{\lambda}(t)$*, representing the "typical" hazard.* Equation (8.11) becomes

$$\boxed{\lambda(t) = \overline{\lambda}(t)e^{a+bU+cE}} \tag{8.14}$$

where now $a = -(b\overline{U} + c\overline{E})$. Equation (8.13) becomes

$$\boxed{\lambda(t) = \overline{\lambda}(t)e^{b(U-\overline{U})+c(E-\overline{E})}} \tag{8.15}$$

Some computer programs for proportional hazard models estimate and print out both the baseline hazard function, $\overline{\lambda}(t)$*, and the coefficients of the predictor variables (in this case, b and c), but they do not print out the constant term a, which must be calculated as* $a = -(b\overline{U} + c\overline{E})$ *in our example.* Because the constant term is not printed out, it is often convenient to express the proportional hazard model in the form given by equation (8.15).

What if we want to use $\lambda'(t)$ instead of $\overline{\lambda}(t)$ as the baseline hazard function? How can we convert the printed values of $\overline{\lambda}(t)$ into values of $\lambda'(t)$ for alternative baseline values $U = U'$ and $E = E'$? From (8.15), the conversion formula is

$$\boxed{\lambda'(t) = \overline{\lambda}(t)e^{b(U'-\overline{U})+c(E'-\overline{E})}} \tag{8.16}$$

The constant term a, to be used if the model is expressed in the form (8.11), is then calculated from (8.9) as $a = -(bU' + cE')$. Once the conversion of the baseline is accomplished, the model can be expressed in the form of either (8.11) or (8.13).

8.1.4. Relative Risks as Measures of Effect on the Hazard

Let us consider the model in the form

$$\lambda(t) = \overline{\lambda}(t)e^{a+bU+cE} \qquad \text{(8.14 repeated)}$$

Suppose that we increase U by one unit (from 0 to 1, representing a change from rural to urban), holding E constant. Denoting the new value of λ as λ^*, we obtain

$$\begin{aligned}\lambda^*(t) &= \bar{\lambda}(t)e^{a+b(U+1)+cE}\\ &= \bar{\lambda}(t)e^{a+bU+cE+b}\\ &= \bar{\lambda}(t)e^{a+bU+cE}e^{b}\\ &= \lambda(t)e^{b}\end{aligned} \tag{8.17}$$

which can be written alternatively as

$$\frac{\lambda^*(t)}{\lambda(t)} = e^b \tag{8.18}$$

From (8.17) it is evident that a one-unit increase in U, holding E constant, multiples the hazard function by e^b. The quantity e^b is called a *relative risk*, for reasons that are obvious from (8.18).

Note that the results in (8.17) *and* (8.18) *do not depend on how the baseline hazard is specified, because both* $\bar{\lambda}(t)$ *and the constant term* a *cancel out of the equations.*

By similar reasoning, one can show that the effect of a one-unit increase in E on the hazard, holding U constant, is e^c.

8.1.5. Hazard Regression Coefficients as Measures of Effect on the Log Hazard

Taking the log of both sides of (8.14) yields

$$\log[\lambda(t)] = \log[\bar{\lambda}(t)] + a + bU + cE \tag{8.19}$$

If we define

$$a'(t) \equiv \log[\bar{\lambda}(t)] + a \tag{8.20}$$

then (8.19) can be rewritten as

$$\log[\lambda(t)] = a'(t) + bU + cE \tag{8.21}$$

This is another way of writing the hazard model. Because (8.21) is in the form of a multiple regression equation, we can immediately interpret b as the effect of a one-unit increase in U on the log hazard, holding E constant. Similarly, we can interpret c as the effect of a one-unit increase in E on the log hazard, holding U constant.

Although proportional hazard models are sometimes presented in log form, the log hazard is not as intuitively meaningful as the hazard itself. *Therefore, e^b (relative risk) is usually preferred over b as the measure of effect.*

It is also instructive to take logs of both sides of (8.15), yielding

$$\log[\lambda(t)] = \log[\bar{\lambda}(t)] + b(U - \bar{U}) + c(E - \bar{E}) \tag{8.22}$$

which can be written

$$\log[\lambda(t)] - \log[\bar{\lambda}(t)] = b(U - \bar{U}) + c(E - \bar{E}) \tag{8.23}$$

Equations (8.21) and (8.23) are analogous to the two forms of an ordinary multiple regression equation with the same predictor variables and, say, overall fertility as the response variable:

$$F = a + bU + cE \tag{8.24}$$

$$F - \bar{F} = b(U - \bar{U}) + c(E - \bar{E}) \tag{8.25}$$

The above discussion may be summarized as follows: *When the* log *hazard is taken as the response variable, effects are additive and the proportional hazard model is viewed as an additive model. When the hazard itself is taken as the response variable, effects are multiplicative and the proportional hazard model is viewed as a multiplicative model.*

Note the similarities with logit regression: Log hazard and hazard are analogous to log odds and odds, and relative risks are analogous to odds ratios. Both relative risks and odds ratios are of the form e^b.

Another point of comparison between logit regression and hazard regression is that the odds ratio approximates a relative risk when the odds become very small. In the odds ratio $[P_1/(1 - P_1)]/[P_2/(1 - P_2)]$, the factors $1 - P_1$ and $1 - P_2$ are then close to unity and approximately cancel, so that the odds ratio then becomes P_1/P_2, which is a relative risk. For example, if $P_1 = .02$ and $P_2 = .01$, then the odds ratio is $(.02/.98)/(.01/.99) \approx .02/.01 = 2$.

8.1.6. The Effect on the Hazard When the Predictor Variable Is Categorical with More Than Two Categories

Let us consider the following set of predictor variables:

M: 1 if medium education, 0 otherwise
H: 1 if high education, 0 otherwise
U: 1 if urban, 0 otherwise
I: 1 if Indian, 0 otherwise

It is convenient to consider the model in the form

$$\lambda(t) = \bar{\lambda}(t)e^{a+bM+cH+dU+fI} \tag{8.26}$$

From (8.26) we can calculate the following values of the hazard $\lambda(t)$ for low, medium, and high education, denoted by $\lambda_L(t)$, $\lambda_M(t)$, $\lambda_H(t)$, respectively:

Low education $(M = 0,\ H = 0)$: $\lambda_L(t) = \bar{\lambda}(t)e^{a+dU+fI}$
Medium education $(M = 1,\ H = 0)$: $\lambda_M(t) = \bar{\lambda}(t)e^{a+b+dU+fI}$
High education $(M = 0,\ H = 1)$: $\lambda_H(t) = \bar{\lambda}(t)e^{a+c+dU+fI}$

Then

$$\frac{\lambda_M(t)}{\lambda_L(t)} = \frac{\bar{\lambda}(t)e^{a+b+dU+fI}}{\bar{\lambda}(t)e^{a+dU+fI}} = e^b \tag{8.27}$$

$$\frac{\lambda_H(t)}{\lambda_L(t)} = \frac{\bar{\lambda}(t)e^{a+c+dU+fI}}{\bar{\lambda}(t)e^{a+dU+fI}} = e^c \tag{8.28}$$

$$\frac{\lambda_M(t)}{\lambda_H(t)} = \frac{\bar{\lambda}(t)e^{a+b+dU+fI}}{\bar{\lambda}(t)e^{a+c+dU+fI}} = e^{b-c} \tag{8.29}$$

the results in (8.27)–(8.29) *do not depend on how the baseline hazard is specified (i.e., the baseline values of* U, M, H, *and* I *need not be chosen as* $\bar{U}$, $\bar{M}$, $\bar{H}$, *and* $\bar{I}$*), because both* $\bar{\lambda}(t)$ *and* a *cancel out of the equations.*

Equations (8.27)–(8.29) can be written

$$\lambda_M(t) = \lambda_L(t)e^b \tag{8.30}$$

$$\lambda_H(t) = \lambda_L(t)e^c \tag{8.31}$$

$$\lambda_M(t) = \lambda_H(t)e^{b-c} \tag{8.32}$$

Equation (8.30) says that the effect of medium education, relative to low education, controlling for U and I, is to multiply the hazard by e^b. Equation (8.31) says that the effect of high education, relative to low education, controlling for U and I, is to multiply the hazard by e^c. Equation (8.32) says that the effect of medium education, relative to high education, is to multiply the hazard by e^{b-c}.

Taking logs of both sides of (8.30)–(8.32), we get

$$\log[\lambda_M(t)] = \log[\lambda_L(t)] + b \tag{8.33}$$

$$\log[\lambda_H(t)] = \log[\lambda_L(t)] + c \tag{8.34}$$

$$\log[\lambda_M(t)] = \log[\lambda_H(t)] + b - c \tag{8.35}$$

Equation (8.33) says that the effect of medium education, relative to low education, controlling for U and I, is to increase (additively) the log hazard by b. Equation (8.34) says that the effect of high education, relative to low education, controlling for U and I, is to increase the log hazard by c. Equation (8.35) says that the effect of medium education, relative to high education, controlling for U and I, is to increase the log hazard by $b - c$.

The reasoning underlying equations (8.26)–(8.35) has a close parallel in logit regression. (See Section 5.2.6 in Chapter 5.)

8.1.7. Interactions

Suppose the model is

$$\lambda(t) = \bar{\lambda}(t)e^{a+bE+cU+dI+fUI} \tag{8.36}$$

where E is once again a quantitative variable, namely, the number of completed years of education. The UI term specifies an interaction between residence and ethnicity. As in ordinary multiple regression, the UI term is treated as a separate variable, which we may think of as $W = UI$, when the model is fitted.

The model is still proportional, because the multiplier $e^{a+bE+cU+dI+fUI}$ *is not a function of time.*

Taking the log of both sides of (8.36), we obtain

$$\log\left[\lambda(t)\right] = \log\left[\bar{\lambda}(t)\right] + a + bE + cU + dI + fUI \qquad (8.37)$$

Aside from the nuisance term $\log[\bar{\lambda}(t)]$, (8.37) is in the form of an ordinary multiple regression equation. We can therefore say immediately that the effect of a one-unit increase in U on $\log \lambda(t)$, holding E and I constant, is to increase $\log \lambda(t)$ by $c + fI$. (See Section 2.6 of Chapter 2.) Denoting the new value of $\lambda(t)$ as $\lambda^*(t)$, we have that

$$\log\left[\lambda^*(t)\right] = \log\left[\lambda(t)\right] + c + fI \qquad (8.38)$$

If we take each side of (8.38) as a power of e, (8.38) becomes

$$\lambda^*(t) = \lambda(t)e^{c+fI} \qquad (8.39)$$

This equation says that the effect of a one-unit increase in U, holding E and I constant, is to multiply the hazard by e^{c+fI}. *In the presence of interaction between* U *and* I, e^{c+fI} *is the relative risk associated with a one-unit increase in* U. Thus the effect of U depends on the level of I, in keeping with the general definition of interaction.

Similarly, a one-unit increase in I, holding E and U constant, increases $\log \lambda(t)$ by $d + fU$ and multiples $\lambda(t)$ by e^{d+fU}.

The reasoning underlying equations (8.36)–(8.39) also has a close parallel in logit regression. (See Section 5.2.8 in Chapter 5.)

Because $a = -(b\bar{E} + c\bar{U} + d\bar{I} + f\overline{UI})$ in (8.36), equations (8.36) and (8.37) can be written alternatively in deviation form as

$$\lambda(t) = \bar{\lambda}(t)e^{b(E-\bar{E})+c(U-\bar{U})+d(I-\bar{I})+f(UI-\overline{UI})} \qquad (8.40)$$

$$\log\left[\lambda(t)\right] = \log\left[\bar{\lambda}(t)\right] + b(E-\bar{E}) + c(U-\bar{U}) + d(I-\bar{I}) + f(UI-\overline{UI}) \qquad (8.41)$$

Because UI is treated as a separate variable, we treat $\overline{UI}$ in these equations as the mean of UI, not as the mean of U times the mean of I.

8.1.8. Nonlinearities

Let us further complicate the model by adding a nonlinearity to the effect that E has on $\lambda(t)$. Let us specify the nonlinearity by adding a quadratic term, E^2:

$$\lambda(t) = \bar{\lambda}(t)e^{a+bE+cE^2+dU+fI+gUI} \tag{8.42}$$

As in ordinary multiple regression, the E^2 term is treated as a separate variable, which we may think of as $W = E^2$, when the model is fitted.

The model is still proportional, because the multiplier $e^{a+bE+cE^2+dU+fI+gUI}$ *is not a function of time.*

Taking logs of both sides of (8.42), we obtain the log form of the model:

$$\log[\lambda(t)] = \log[\bar{\lambda}(t)] + a + bE + cE^2 + dU + fI + gUI \tag{8.43}$$

Because (8.43) is in the form of a multiple regression equation, we can immediately say that the (instantaneous) effect of a one-unit increase in E is to increase $\log[\lambda(t)]$ by $b + 2cE$. (See Section 2.7.1 in Chapter 2.) Denoting the new value of $\lambda(t)$ by $\lambda^*(t)$, we have that

$$\log[\lambda^*(t)] = \log[\lambda(t)] + b + 2cE \tag{8.44}$$

Taking both sides of (8.44) as a power of e, we get

$$\lambda^*(t) = \lambda(t)e^{b+2cE} \tag{8.45}$$

Thus the relative risk associated with a one-unit increase in E is e^{b+2cE}. The effect of e depends on the level of E, in keeping with the general definition of nonlinearity.

These results also have a close parallel in logit regression. (See Section 5.2.9 in Chapter 5.)

Equations (8.42) and (8.43) can be written alternatively in deviation form as

$$\lambda(t) = \bar{\lambda}(t)e^{b(E-\bar{E})+c(E^2-\overline{E^2})+d(U-\bar{U})+f(I-\bar{I})+g(UI-\overline{UI})} \tag{8.46}$$

$$\log[\lambda(t)] = \log[\bar{\lambda}(t)] + b(E - \bar{E}) + c\left(E^2 - \overline{E^2}\right) + d(U - \bar{U}) + f(I - \bar{I}) + g\left(UI - \overline{UI}\right) \tag{8.47}$$

Because E^2 is treated as a separate variable, we treat $\overline{E^2}$ as the mean of E^2, not the mean of E times the mean of E.

8.1.9. Changing the Reference Category of a Categorical Predictor Variable

This is done in the same way as described in Chapter 5 for logit regression. A change in reference category does not require a rerun of the model.

8.2. CALCULATION OF LIFE TABLES FROM THE PROPORTIONAL HAZARD MODEL

8.2.1. The Hazard Model in Survivorship Form

In Chapter 7, we derived the formula

$$\log[S(t)] = -\int_0^t \lambda(x)\,dx \qquad \text{(7.26 repeated)}$$

To illustrate how this formula can be used to express the hazard model in survivorship form, let us reconsider the simple model

$$\lambda(t) = \overline{\lambda}(t)e^{a+bU+cE} \qquad \text{(8.14 repeated)}$$

Substituting (8.14) into (7.26), we obtain

$$\begin{aligned}\log[S(t)] &= \int_0^t \overline{\lambda}(x)e^{a+bU+cE}\,dx \\ &= \left[-\int_0^t \overline{\lambda}(x)\,dx\right]e^{a+bU+cE} \\ &= \left(\log\left[\overline{S}(t)\right]\right)e^{a+bU+cE} \end{aligned} \qquad (8.48)$$

where $\overline{S}(t)$ is the survivorship function for persons for whom $U = \overline{U}$ and $E = \overline{E}$. The last step in the derivation of (8.48) makes use of (7.26) repeated above. *Note that we can factor the term $e^{a+bU+cE}$ out of the integral only because the model is proportional*—that is, only because the term $e^{a+bU+cE}$ does not depend on t. [Recall the definition of proportionality in equation (8.2).]

Because $a \log b = \log[b^a]$, we can rewrite (8.48) as

$$\log[S(t)] = \log\left\{\left[\overline{S}(t)\right]^{e^{a+bU+cE}}\right\} \qquad (8.49)$$

Taking each side of (8.49) as a power of e, we obtain

$$S(t) = \left[\bar{S}(t)\right]^{e^{a+bU+cE}} \tag{8.50}$$

Alternatively, in deviation form [because $a = -(b\bar{U} + c\bar{E})$]

$$\boxed{S(t) = \left[\bar{S}(t)\right]^{e^{b(U-\bar{U})+c(E-\bar{E})}}} \tag{8.51}$$

which is more convenient for calculation. Note that there are two levels of exponentiation in (8.51). One first calculates $e^{b(U-\bar{U})+c(E-\bar{E})}$. Denoting this value as A, one then calculates $S(t)$ as $[\bar{S}(t)]^A$. Alternatively, if one has a calculator without a y^x key, one can start by calculating the right side of (8.48); denoting the result of this calculation by B, one then calculates $S(t)$ as e^B.

When the proportional hazard model in (8.14) is fitted to the data, values of the means, $\bar{U}$ and $\bar{E}$, and the estimated coefficients, a, b and c, are printed out. The estimated survival function $\bar{S}(t)$ is also printed out. Suppose that we wish to estimate the probability $F(120)$ of having a sixth birth by $t = 120$ for urban persons with 5 years of education. We substitute $t = 120$, $U = 1$, and $E = 5$ into (8.51), yielding an estimated value of $S(120)$. We then obtain $F(120) = 1 - S(120)$.

8.2.2. The Problem of Unobserved Heterogeneity

Suppose that we have fitted the model $\lambda(t) = \bar{\lambda}(t)e^{b(E-\bar{E})}$ and found that $b < 0$ (the higher the education, the lower the hazard of a sixth birth). Let us assume that the model fits the data very closely. How does the life table based on $\bar{\lambda}(t)$, obtained from this model, compare with an ordinary product-limit life table calculated for the same sample? Bear in mind that the hazard model, as specified, allows for heterogeneity with respect to education in the hazard of a sixth birth, whereas the ordinary product-limit life table erroneously assumes homogeneity (i.e., it assumes that each sample member has the same hazard function).

Let us denote life-table quantities for the ordinary product-limit life table by a subscript P. The problem with $S_P(t)$ is that the homogeneity assumption is violated, insofar as, over time, less educated women get thinned out of the life-table cohort faster than more educated women, because the hazard of a sixth birth is lower the higher the education. In other words, the educational composition of the cohort changes over time. In contrast, in the case of $\bar{S}(t)$, education is held constant at $\bar{E}$ as the cohort moves forward in time. Therefore, $S_P(t)$ and $\bar{S}(t)$ will generally differ to some extent.

Survival models never include enough predictor variables to account fully for heterogeneity. Product-limit life tables do not allow at all for heterogeneity, and proportional hazard models capture only some of it. In proportional hazard models, there is always some remaining unobserved heterogeneity that biases the results to an unknown extent.

As in ordinary multiple regression and logit regression, the coefficients of the predictor variables may also be biased, due to our inability to control for all sources of variation.

8.2.3. Test for Proportionality

Suppose that we have urban and rural life tables, the hazard functions of which are proportional. Denoting the two observed life tables by subscripts u (urban) and r (rural), we can represent this proportionality as $\lambda_u(t)/\lambda_r(t) = c$, where c is some positive constant. Because a positive constant c can be written as e^k for $k = \log c$, we can write alternatively $\lambda_u(t)/\lambda_r(t) = e^k$ or, equivalently,

$$\lambda_u(t) = \lambda_r(t)e^k \tag{8.52}$$

From equation (7.26) in Chapter 7, we also have

$$\log[S_u(t)] = -\int_0^t \lambda_u(x)\,dx \tag{8.53}$$

Substituting (8.52) into (8.53), we obtain

$$\begin{aligned}\log[S_u(t)] &= -\int_0^t \lambda_r(x)e^k\,dx \\ &= \left[-\int_0^t \lambda_r(x)\,dx\right]e^k \\ &= (\log[S_r(t)])e^k \end{aligned} \tag{8.54}$$

Note that we can factor the term e^k out of the integral only because the model is proportional. Multiplying both sides of (8.54) by -1, taking logs of both sides again, and rearranging terms, we obtain

$$\boxed{\log[-\log S_u(t)] - \log[-\log S_r(t)] = k} \tag{8.55}$$

Equation (8.55) suggests a simple graphical approach for testing whether the urban–rural residence variable is acting proportionally on the hazard. *All we need to do is plot values of* $\log[-\log S_u(t)] - \log[-\log S_r(t)]$ *against* t. *If the plot approximates a horizontal line* (*i.e., a constant*), *the proportionality assumption is warranted.* If the plot deviates systematically from a horizontal line (e.g., if the plot approximates a line with a substantially positive or negative slope), the proportionality assumption is not warranted.

An even simpler test of proportionality would be to plot $\lambda_u(t)/\lambda_r(t)$ itself against t. If the hazard rates are proportional, the plot will approximate a horizontal line. The reason why this plot is not usually used is that individual hazard rates (for, say, one-month intervals) are often based on very few failures, so that the hazard rates fluctuate a great deal due to sampling variability. The cumulative survival curve, $S(t)$, is based on many more failures, so that the plot of $\log[-\log S_u(t)] - \log[-\log S_r(t)]$ is much smoother than the plot of $\lambda_u(t)/\lambda_r(t)$, especially at higher values of t.

When the model has more than one predictor variable, the test for proportionality is more complicated. For example, if the predictor variables include education as well as urban–rural residence, a test of the proportionality of the effects of urban–rural residence requires holding education constant. This can be done, for example, by grouping educational levels into categories and then testing for the proportionality or urban–rural residence effects for each category of education separately.

Unfortunately, this strategy breaks down when more predictor variables are added to the model. The predictor variables must be cross-classified into categories that rapidly become numerous as more predictor variables are added. The number of separate life tables to be calculated becomes large and we quickly run out of cases.

In practice what we do is a series of very rough tests of proportionality. For example, we test whether residence behaves proportionally by plotting $\log[-\log S_u(t)] - \log[-\log S_r(t)]$ against t without controlling for education, and we test whether education behaves proportionally by plotting, say, $\log[-\log S_{low}(t)] - \log[-\log S_{high}(t)]$ against t without controlling for residence (subscripts *low* and *high* denote low and high education, respectively). If the plots are approximately horizontal, they provide a rough indication that the proportionality assumption may be acceptable.

Of course, we can always impose the proportionality assumption on the data and run the model anyway. But the result may be a poor fit, as indicated by high standard errors and statistical nonsignificance of coefficients.

8.3. STATISTICAL INFERENCE AND GOODNESS OF FIT

As already mentioned, hazard models are fitted by the maximum likelihood method. The similarity to logit regression has already been noted. Tests of coefficients, tests of the difference between two coefficients, and tests of the difference between two models are done in the same way as in logit regression.

Pseudo-R may be calculated as a measure of goodness of fit in the same way as in logit regression, but it suffers the same drawbacks already mentioned in Chapters 5 and 6. Pseudo-R is seldom used in the context of hazard models.

8.4. A NUMERICAL EXAMPLE

Let us consider the model

$$\lambda(t) = \overline{\lambda}(t)e^{a+bU+cH+dUH+fA+gA^2+hI} \tag{8.56}$$

where

$\lambda(t)$ = estimated hazard of a sixth birth

U = 1 if urban, 0 otherwise

H = 1 if high education (including medium and high as defined previously), 0 otherwise

A = woman's age at time of fifth birth

I = 1 if Indian, 0 otherwise

High and medium education, as defined previously, are combined into a new "high" category because, in Fiji, very few women who had a fifth birth were in the original "high" category of 8 or more completed years of education. The new "high" category includes women with 5 or more completed years of education. A UH term is included to allow for the possibility that education may have a larger effect on the hazard of a sixth birth in urban areas than in rural areas.

The first step, before fitting the model, is to test for proportionality. We begin by examining whether the effects of residence are proportional, without controlling for other variables. Figure 8.2a plots product-limit life-table estimates of $S(t)$ by residence, based on separate life tables for urban and rural, and Figure 8.2b plots the urban–rural difference in $\log[-\log S(t)]$. In the birth histories, dates of birth were coded to the nearest month, so at most there is one plotted point per month.

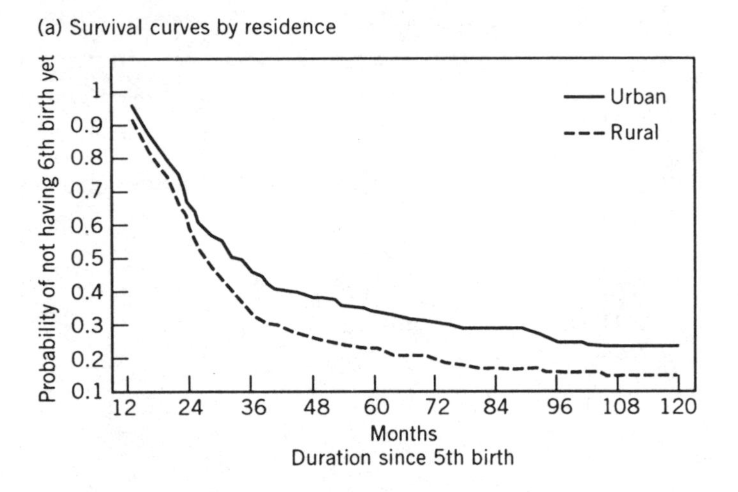

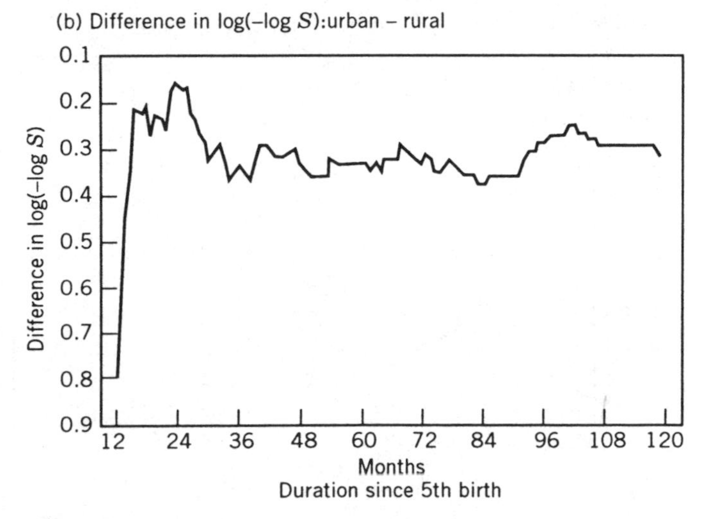

Figure 8.2. Graphical test of proportionality of residence effects.

In Figure 8.2b, the plot of the urban–rural difference in $\log[-\log S(t)]$ conforms fairly well to a horizontal line, except for three points at very short durations that are based on a handful of cases. We conclude that the proportionality assumption is acceptable for residence effects.

Similar plots (not shown) for education and age also indicate that the proportionality assumption is acceptable for these variables. The test of whether age acts proportionally involves the categorization of age into broad age groups.

Ethnicity, on the other hand, does not act proportionally. Figure 8.3a shows that the survival curves for Indians and Fijians cross over

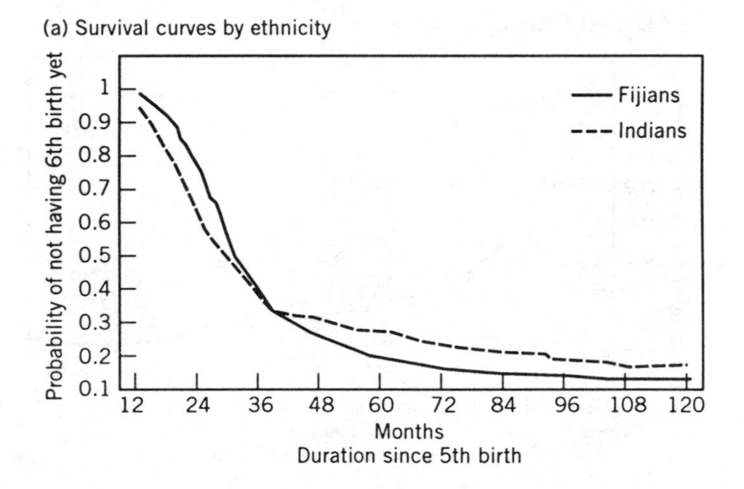

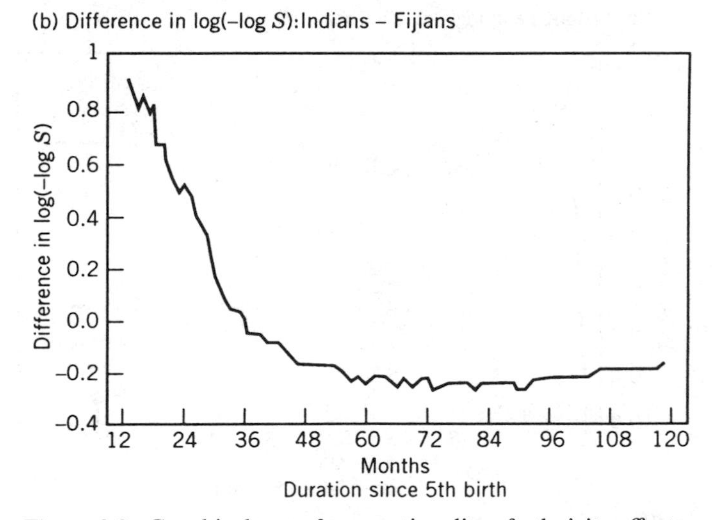

Figure 8.3. Graphical test of proportionality of ethnicity effects.

each other. Indians evidently have higher hazard rates than Fijians at short durations but have lower hazard rates at higher durations. The survival curve crosses over at about 36 months, at which time it is about .4. This translates into a ratio of progression from fifth to sixth birth, $F(36)$, of $1 - .4 = .6$. The ultimate progression ratio is higher for Fijians than for Indians. At 120 months the progression ratio is $1 - .18 = .82$ for Indians and $1 - .13 = .87$ for Fijians.

The crossover of survival curves is reflected in a major departure from horizontal linearity in the plot of the Indian–Fijian difference in $\log[-\log S(t)]$ in Figure 8.3b. We conclude that ethnicity does not behave proportionally. We accordingly revise the model by dropping ethnicity as a predictor variable:

$$\lambda(t) = \bar{\lambda}(t)e^{a+bU+cH+dUH+fA+gA^2} \tag{8.57}$$

A possible next step is to run this simplified model separately for Indians and Fijians.

For Fijians, no coefficients or effects are statistically significant. In the interest of brevity, we present results only for Indians. At first sight, it appears from Table 8.1 that there are no significant effects for Indians either. Z scores, calculated by dividing each coefficient by its standard error, are all less than 1.96. *When there is interaction or*

TABLE 8.1. Coefficients and Multiplicative Effects for the Model $\lambda(t) = \bar{\lambda}(t)e^{a+bU+cH+dUH+fA+gA^2}$, Where $\lambda(t)$ Is the Estimated Hazard of Having a Sixth Birth: Indian Women Who Have Had at least Five Births, 1974 Fiji Fertility Survey[a]

Predictor Variable	Coefficient	Effect
Residence × education		
U	$-.1692(.0928)$	$e^{-.1692-.2507H}$
H	$-.0898(.1153)$	$e^{-.0898-.2507U}$
UH	$-.2507(.1868)$	—
Age		$e^{.0412+2(-.0022)A}$
A	$.0412(.0935)$	—
A^2	$-.0022(.0017)$	—

[a]Numbers in parentheses are standard errors of coefficients. None of the coefficients are significant at the 5 percent level. (Z scores, obtained by dividing each coefficient by its standard error, are all less than 1.96.) The entry for the effect of A uses the formula for the instantaneous effect of A. Based on the log likelihood ratio test, $p = .0000$ to four decimal places, indicating that the model differs very significantly from the intercept model.

nonlinearity, however, coefficients are not the same as effects (see Sections 2.9.3 and 2.9.4 in Chapter 2). We must test the significance of effects as well as the significance of coefficients.

In the case of the interaction between residence and education, we must test differences between cross-classified categories of residence (rural, urban) and education (low, high). We have that

$$\begin{aligned}
&\text{Rural, low} \quad (U=0, H=0): \quad \lambda(t) = \bar{\lambda}(t)e^{a+fA+gA^2} \\
&\text{Rural, high} \quad (U=0, H=1): \quad \lambda(t) = \bar{\lambda}(t)e^{a+c+fA+gA^2} \\
&\text{Urban, low} \quad (U=1, H=0): \quad \lambda(t) = \bar{\lambda}(t)e^{a+b+fA+gA^2} \\
&\text{Urban, high} \quad (U=1, H=1): \quad \lambda(t) = \bar{\lambda}(t)e^{a+b+c+d+fA+gA^2}
\end{aligned}$$

Judging from these formulae and the values of the coefficients b, c, and d, which are all negative (Table 8.1), the largest relative difference is between urban-high and rural-low:

$$\begin{aligned}
\frac{\lambda_{U,H}(t)}{\lambda_{R,L}(t)} = \frac{\bar{\lambda}(t)e^{a+b+c+d+fA+gA^2}}{\bar{\lambda}(t)e^{a+fA+gA^2}} &= e^{b+c+d} \\
&= e^{-.1692-.0898-.2507} \\
&= .60
\end{aligned} \tag{8.58}$$

Equation (8.58) says that the effect of being urban-high, relative to rural-low, is to reduce the hazard by 40 percent [(1 − .60) × 100].

Is this effect statistically significant? To address this question, we must test whether e^{b+c+d} differs significantly from one, or, equivalently, whether $b + c + d$ differs significantly from zero.

For this latter test, we can calculate a Z score as[1]

$$\begin{aligned}
Z &= \frac{b+c+d}{s_{b+c+d}} = \frac{b+c+d}{\sqrt{\operatorname{Var}(b+c+d)}} \\
&= \frac{b+c+d}{\sqrt{\operatorname{Var}(b)+\operatorname{Var}(c)+\operatorname{Var}(d)+2\operatorname{Cov}(b,c)+2\operatorname{Cov}(b,d)+2\operatorname{Cov}(c,d)}}
\end{aligned} \tag{8.59}$$

[1] In (8.63), we make use of the general formula

$$\operatorname{Var}\left(\sum_i a_i X_i\right) = \sum_i a_i^2 \operatorname{Var}(X_i) + 2\sum_{j>i} a_i a_j \operatorname{Cov}(X_i, X_j)$$

The covariances are obtained from the regression coefficient covariance matrix (labeled "estimated asymptotic covariance matrix" in BMDP), which can be printed out by most computer programs as an option. The variances are obtained either from the diagonal terms of the regression coefficient covariance matrix or, as shown in (8.60) below, by squaring the standard errors of b, c, and d. With numbers substituted in, (8.59) becomes

$$Z = \frac{-.1692 - .0898 - .2507}{\sqrt{(.0928)^2 + (.1153)^2 + (.1868)^2 + 2(.0026) + 2(-.0086) + 2(-.0133)}}$$

$$= -3.8 \tag{8.60}$$

The numbers in this equation come from Table 8.1, except for the covariances, which are taken from the regression coefficient covariance matrix (not shown). Because $|Z| = 3.8 > 1.96$, we conclude that the effect of being urban-high, relative to rural-low, on the hazard is statistically significant at the 5 percent level.

Similarly, the effect of being urban-high, relative to urban-low, on the hazard is

$$\frac{\lambda_{U,H}(t)}{\lambda_{U,L}(t)} = e^{c+d} = e^{-.0898-.2507} = .71 \tag{8.61}$$

To test the statistical significance of this effect, we calculate the Z score:

$$Z = \frac{c+d}{s_{c+d}} = \frac{c+d}{\sqrt{\text{Var}(c+d)}}$$

$$= \frac{c+d}{\sqrt{\text{Var}(c) + \text{Var}(d) + 2\,\text{Cov}(c,d)}}$$

$$= \frac{-.0898 - .2507}{\sqrt{(.1153)^2 + (.1868)^2 + 2(.0133)}} = -2.3 \tag{8.62}$$

Thus, from (8.61), the effect of high education, relative to low education, among urban residents is to reduce the hazard by 29 percent [(1 − .71) × 100]. Becausc $|Z| - 2.3 > 1.96$, this effect is also significant at the 5 percent level.

It can similarly be shown that the effect of being urban-high, relative to rural-high, is to reduce the hazard by 34 percent, with $Z = -2.6$. This effect is also significant at the 5 percent level.

The remaining effects are nonsignificant.

The above example is especially instructive, because it shows that *effects can be significant even when the coefficients that underly them are not*. The lesson is clear: When there is interaction between two categorical variables, one must first cross-classify the categories; then, for each pair of cross-classified categories, one must test whether the effect of being in one category, relative to the other, on the hazard is statistically significant.

It is also instructive to test the significance of the effect of a one-unit increase in A on the hazard, evaluated at different values of A. At age 25, for example, this effect is to multiply the hazard by $e^{f+2gA} = e^{f+50g} = e^{.0412+50(-.0022)} = .93$. Thus, at $A = 25$, the effect of a one-unit increase in A is to decrease the hazard by 7 percent.

To test whether this effect is significant, we compute

$$Z = \frac{f + 50g}{s_{f+50g}} = \frac{f + 50g}{\sqrt{\text{Var}(f + 50g)}}$$

$$= \frac{f + 50g}{\sqrt{\text{Var}(f) + (50)^2\,\text{Var}(g) + 100\,\text{Cov}(f, g)}} \tag{8.63}$$

where the formula for $\text{Var}(f + 50g)$ is derived from the general formula given earlier in footnote 1.

Evaluating (8.63) from the BMDP output turned out to be problem, because the estimates of standard errors, variances, and covariances, some of which are very small, are printed only to four decimal places, resulting in excessive rounding. For example, the regression coefficient covariance matrix yields $\text{Cov}(f, g) = -.0002$, which is a highly rounded figure with only one significant digit. (Leading zeros are not counted as significant digits; note that in this context, "significant" does not mean "statistically significant.") The matrix also specifies $\text{Var}(g)$ as .0000, which is so highly rounded that no significant digits are shown. One can calculate $\text{Var}(g)$ alternatively by squaring the standard error of g, .0017, but this figure also is rounded to only two significant digits. We discovered that rounding was a serious problem when our computation of the quantity under the square root sign in (8.63) yielded a negative number, which is impossible.

To get around the problem of excessive rounding, we created a new age variable, $A' = A/10$, and reran the model with A' in place of A. A one-unit change in A' from, say, 0 to 1, is equivalent to a 10-unit change in A, from 0 to 10. Because $(A')^2 = A^2/100$, a one-unit change in $(A')^2$ from, say, 0 to 1, is equivalent to a 100-unit change in A^2 from 0 to 100. Therefore the coefficient f' must be 10 times larger than f, and g' must be 100 times larger than g. This trick gets rid of some leading zeros in the estimates of standard errors, variances, and covariances and therefore reveals more significant digits. One then gets the coefficient f by dividing the coefficient f' by 10, and the coefficient g by dividing the coefficient g' by 100.

With these more precise values, we can calculate Z from (8.63) as

$$Z = \frac{.04119 + 50(-.00218)}{\sqrt{(.09351)^2 + (50)^2(.001749)^2 + 100(-.000163)}} = -7.1 \tag{8.64}$$

Because rounding errors can be serious in this kind of calculation, results must be checked carefully.

One can similarly calculate that the effect of a one-unit increase in A is to reduce the hazard by 9 percent at $A = 30$ ($Z = -6.4$), 11 percent at $A = 35$ ($Z = -3.7$), and 12 percent at $A = 40$ ($Z = -2.8$). *In every case* $|Z| > 1.96$; *therefore, all of these effects are significant at the* 5 *percent level, even though the underlying coefficients of* A *and* A^2 *are not significant.*

8.5. MULTIPLE CLASSIFICATION ANALYSIS (MCA) ADAPTED TO PROPORTIONAL HAZARD REGRESSION

Let us continue with the model

$$\lambda(t) = \bar{\lambda} e^{a + bU + cH + dUH + fA + gA^2} \tag{8.57 repeated}$$

fitted to Indians.

In hazard regression, the MCA table is usually constructed for a single life-table parameter. Suppose that the parameter of interest is $F(120)$, the proportion having a sixth birth within 120 months of having a fifth birth. Using the deviation form of the model, we have

$$\begin{aligned} F(120) &= 1 - S(120) \\ &= 1 - \bar{S}(120) e^{b(U - \bar{U}) + c(H - \bar{H}) + d(UH - \overline{UH}) + f(A - \bar{A}) + g(A^2 - \overline{A^2})} \end{aligned} \tag{8.65}$$

TABLE 8.2. MCA Table of Adjusted Values of $F(120)$ for the Model $\lambda(t) = \bar{\lambda}(t)e^{a+bU+cH+dUH+fA+gA^2}$, Where $\lambda(t)$ Is the Estimated Hazard Having a Sixth Birth: Indian Women Who Have Had at Least Five Births, 1974 Fiji Fertility Survey[a]

Predictor Variable	n	Adjusted Value of $F(120)$	
		Formula	Numerical Value
Residence × education			
Rural, low	493	$1-[\bar{S}(120)]^{e^{-b\bar{U}-c\bar{H}-d\overline{UH}}}$	.85
Rural, high	145	$1-[\bar{S}(120)]^{e^{-b\bar{U}+c(1-\bar{H})-d\overline{UH}}}$	.83
Urban, low	226	$1-[\bar{S}(120)]^{e^{b(1-\bar{U})-cH-d\overline{UH}}}$	.80
Urban, high	117	$1-[\bar{S}(120)]^{e^{b(1-\bar{U})+c(1-\bar{H})+d(1-\overline{UH})}}$	.68
Age at fifth birth			
25	—	$1-[\bar{S}(120)]^{e^{f(25-\bar{A})+g(625-\overline{A^2})}}$	.86
30	—	$1-[\bar{S}(120)]^{e^{f(30-\bar{A})+g(900-\overline{A^2})}}$	.73
35	—	$1-[\bar{S}(120)]^{e^{f(35-\bar{A})+g(1225-\overline{A^2})}}$	.55
40	—	$1-[\bar{S}(120)]^{e^{f(40-\bar{A})+g(1600-\overline{A^2})}}$	.35
Total	981		

[a]Fitted values of the coefficients b, c, d, f, and g are given in Table 8.1. Mean values of the predictor variables are $\bar{U} = .3496$, $\bar{H} = .2671$, $\overline{UH} = .1193$, $\bar{A} = 26.4965$, and $\overline{A^2} = 719.3826$. Also $\bar{S}(120) = .1816$. Substitution of these values into the formulae given in the table yields the numerical values shown in the last column. Calculation formulae are based on equation (8.65) in the text.

This formula, which is similar in form to equation (8.51) derived earlier, provides the basis for calculating adjusted values of $F(120)$. Calculation formulae and results are summarized in Table 8.2. As usual, when we look at the effect of one predictor variable, we control for the other predictor variables by setting them at their means in the entire sample.

8.6. FURTHER READING

An easy-to-read introduction to proportional hazard models may be found in Allison (1984). For more advanced treatments, see Kalbfleisch and Prentice (1980) and Cox and Oakes (1984).

9

SURVIVAL MODELS, PART 3: HAZARD MODELS WITH TIME DEPENDENCE

In proportional hazard models, considered in Chapter 8, predictor variables and coefficients of predictor variables are not functions of time. We now consider hazard models where predictor variables or coefficients of predictor variables depend on time. Because the term "proportional" is no longer appropriate, we refer to this class of survival models as *hazard models with time dependence*.

9.1. TIME-DEPENDENT PREDICTOR VARIABLES

Let us consider the following model, fitted to Indian women who have had at least five births, as reported in the 1974 Fiji Fertility Survey:

$$\lambda(t) = \bar{\lambda}(t)e^{a+bU+cH+dUH+fA+gA^2+hD(t)} \tag{9.1}$$

where

$\lambda(t)$ = estimated hazard of a sixth birth (starting event is fifth birth)

U = 1 if urban, 0 otherwise

H = 1 if five or more completed years of education, 0 otherwise

A = woman's age at time of fifth birth

$D(t)$ = 1 if one of the woman's first five children died twelve or more months ago, 0 otherwise

As in previous chapters, the model is stated in estimated form. Although we have chosen to fit the model to Indians, we could just as well have chosen Fijians.

This model is the same as the model considered earlier in Tables 8.1 and 8.2 and in Figures 8.2 and 8.3, except that $D(t)$ has been added to the model. Clearly, for a particular woman, $D(t)$ is not necessarily constant over time. $D(t)$ could be zero for some months following a fifth birth, then change to one due to the death of a prior child. Because $D(t)$ may change over time, it is called a *time-dependent predictor variable*. The reason for including $D(t)$ in the model is that the loss of one of a woman's first five children may cause her to go on to have a sixth child. Because it usually takes at least three months to conceive plus nine months of pregnancy before a birth arrives, we have specified a 12-month time lag in the definition of $D(t)$.

The hazard model in (9.1) is fitted using the maximum likelihood method. As usual, we shall not be concerned with the details of how this is done, except to note that fitting a hazard model with time dependence requires considerably more computer time than fitting a proportional hazard model.

9.1.1. Coefficients and Effects

The effect of a one-unit increase in D, holding U, H, A, and t constant, is to multiply the hazard by e^h, the same as would be the case if D were not time-dependent. The mathematical logic used to derive this result is also the same. (See Section 8.1.4) in Chapter 8.)

Table 9.1 shows fitted coefficients and estimated effects for the model in equation (9.1). Table 9.1 has the same format as Table 8.1, except that $D(t)$ has been added to the model. The coefficients and effects of variables other than $D(t)$ turn out to be about the same as they were in Table 8.1 for the model without $D(t)$, indicating that the effects of residence, education, and age on the hazard of a sixth birth are not appreciably influenced by the death of an earlier child. Statistical tests of significance of coefficients and effects of these variables are done in the same way as in the proportional hazard models considered in Chapter 8 and are not considered any further

TABLE 9.1. Coefficients and Multiplicative Effects for the Model $\lambda(t) = \bar{\lambda}(t)e^{a + bU + cH + dUH + fA + gA^2 + hD(t)}$, Where $\lambda(t)$ Is the Estimated Hazard of Having a Sixth Birth: Indian Women Who Have Had at Least Five Births, 1974 Fiji Fertility Survey[a]

Predictor Variable	Coefficient	Effect
Residence × education		
U	−.1471 (.0932)	$e^{-.1471-.2775H}$
H	−.0733 (.1154)	$e^{-.0733-.2775U}$
UH	−.2775 (.1872)	—
Age		$e^{.0560+2(-.0024)A}$
A	.0560 (.0932)	—
A^2	−.0024 (.0017)	—
Child death		
$D(t)$	.2043* (.0808)	$e^{.2043} = 1.23$

[a]Numbers in parentheses are standard errors of coefficients. An asterisk after a coefficient indicates that it is significant at the 5 percent level.

here. The coefficient of the added variable, $D(t)$, is .2043, which is significant at the 5 percent level ($Z = .2043/.0808 = 2.53 > 1.96$). As shown in the column labeled "effect," the death of an earlier child increases the hazard of a sixth birth by 23 percent, holding constant the other variables in the model.

It is evident from this example that the ability to handle time-dependent predictor variables is an attractive feature of hazard models. Note the ease by which a causal ordering is established in the case of previous child death. The causal ordering is simply a time ordering: A child death at time $t - 12$ or earlier has an effect on the hazard at time t. To be sure, one must have longitudinal or retrospective data at one's disposal in order to establish the time ordering.

9.1.2. Choosing a Baseline Hazard Function

Recall that, for proportional hazard models, the output contains not only fitted values of the coefficients of the predictor variables, but also a fitted baseline hazard function, which is specified at the mean values of the predictor variables. The fitting is accomplished in two steps that were not described in Chapter 8: First, the coefficients are estimated by means of a maximum likelihood function that takes the form of a *partial likelihood function*. The term "partial" is used because the baseline hazard function cancels out of the likelihood function, leaving only the coefficients to be estimated. Second, the fitted coefficients are substituted into a second estimation procedure that yields a fitted baseline hazard function specified at the mean values of the predictor variables.

Only the first of these two steps works in the case of hazard models with time dependence; in other words, we are able to estimate the coefficients but not the baseline hazard function. This is fine if all we are interested in is the estimation of effects on the hazard. But what if we want to present results in the form of a multiple classification analysis (MCA) table using a measure like $F(120) = 1 - S(120)$? To calculate $F(120)$, we need to calculate the survivorship function $S(t)$, which is calculated from $\lambda(t)$, which is in turn calculated from the estimated coefficients and the baseline hazard $\overline{\lambda}(t)$.

We must therefore make a rough estimate of the baseline hazard function. The way this is done will depend to some extent on the particular hazard model being investigated and on the aims of analysis. In Section 9.1.3 we will illustrate how appropriate baseline hazard functions can be chosen for the model in (9.1), in the context of constructing an MCA table similar to that shown earlier in Table 8.2.

9.1.3. MCA Adapted to Hazard Regression with Time-Dependent Predictor Variables

We would like to construct a baseline hazard function specified at $\overline{U}$, $\overline{H}$, $\overline{UH}$, $\overline{A}$, $\overline{A^2}$, and $\overline{D}(t)$. The best we can do is to ignore the time-dependent predictor variable $D(t)$, fit the resulting *proportional* hazard model

$$\lambda(t) = \overline{\lambda}(t)e^{a+bU+cH+dUH+fA+gA^2} \tag{9.2}$$

and use the baseline hazard function from this proportional hazard model. This baseline hazard, which is specified at $\overline{U}$, $\overline{H}$, $\overline{UH}$, $\overline{A}$, and $\overline{A^2}$, will in general be slightly biased, insofar as child deaths, being statistically uncontrolled, are a source of unobserved heterogeneity. The baseline survivorship function $\overline{S}(t)$, which is derived from the baseline hazard function $\overline{\lambda}(t)$, will also be slightly biased.

We can use $\overline{S}(t)$, which is printed out for the truncated model in (9.2), to calculate the residence $\times$ education panel and the age panel of the MCA table for the full model in equation (9.1), as shown in Table 9.2. *The formulae in these two panels use $\overline{S}(120)$ derived from the truncated model in (9.2), but coefficients b, c, d, f, and g derived from*

TABLE 9.2. MCA Table of Adjusted Values of $F(120)$ for the Model $\lambda(t) = \bar{\lambda}(t)e^{a+bU+cH+dUH+fA+gA^2+hD(t)}$, Where $\lambda(t)$ Is the Estimated Hazard of Having a Sixth Birth: Indian Women Who Have Had at Least Five Births, 1974 Fiji Fertility Survey[a]

Predictor Variable	n	Adjusted Value of $F(120)$: Formula	Numerical Value
Residence × education			
Rural, low	493	$1 - [\bar{S}(120)]^{e^{-b\bar{U}-c\bar{H}-d\overline{UH}}}$	.85
Rural, high	145	$1 - [\bar{S}(120)]^{e^{-b\bar{U}+c(1-\bar{H})-d\overline{UH}}}$	.83
Urban, low	226	$1 - [\bar{S}(120)]^{e^{b(1-\bar{U})-c\bar{H}-d\overline{UH}}}$	.80
Urban, high	117	$1 - [\bar{S}(120)]^{e^{b(1-\bar{U})+c(1-\bar{H})+d(1-\overline{UH})}}$	.68
Age			
25	—	$1 - [\bar{S}(120)]^{e^{f(25-\bar{A})+g(625-\overline{A^2})}}$	.84
30	—	$1 - [\bar{S}(120)]^{e^{f(30-\bar{A})+g(900-\overline{A^2})}}$	.72
35	—	$1 - [\bar{S}(120)]^{e^{f(35-\bar{A})+g(1225-\overline{A^2})}}$	.54
40	—	$1 - [\bar{S}(120)]^{e^{f(40-\bar{A})+g(1600-\overline{A^2})}}$	.34
Child death			
At $t = -12$	—	$1 - \exp\left[-\sum_{t=0}^{119} \lambda'_t e^h\right]$	.85
At $t = 0$	—	$1 - \exp\left[-\left(\sum_{t=0}^{11} \lambda'_t + \sum_{t=12}^{119} \lambda'_t e^h\right)\right]$	.85
At $t = 24$	—	$1 - \exp\left[-\left(\sum_{t=0}^{35} \lambda'_t + \sum_{t=36}^{119} \lambda'_t e^h\right)\right]$	.82
At $t = 48$	—	$1 - \exp\left[-\left(\sum_{t=0}^{59} \lambda'_t + \sum_{t=60}^{119} \lambda'_t e^h\right)\right]$	.81
At $t = 60$	—	$1 - \exp\left[-\left(\sum_{t=0}^{71} \lambda'_t + \sum_{t=72}^{119} \lambda'_t e^h\right)\right]$	.80
No child dies	—	$1 - \exp\left[-\sum_{t=0}^{119} \lambda'_t\right]$	.79

[a]See footnote to Table 8.2 for the means of U, H, UH, A, and A^2 and the value of $\bar{S}(120)$. See Table 9.1 for values of the coefficients b, c, d, f, g, and h. Since values of λ'_t are not shown, the reader cannot replicate the numerical values in the child death panel.

the full model in (9.1). The formulae (which are in deviation form) are the same as those in Table 8.2 in Chapter 8, except for the values assigned to the coefficients.

Because $D(t)$ is time-dependent, the mode of calculation is different for the child death panel in Table 9.2. One possible way to proceed is to choose the baseline hazard for the child death panel of Table 9.2 as the baseline hazard $\lambda'(t)$ for the *proportional* hazard model

$$\lambda(t) = \lambda'(t)e^{a+bU+cH+dUH+fA+gA^2} \tag{9.3}$$

fitted to the subset of Indian women who had a fifth birth but no child deaths before having a sixth birth—that is, to the subset of Indian women for whom $D(t) = 0$ *for every* t *up to sixth birth or censoring, whichever comes first*. Baseline values of the predictor variables, for purposes of calculating $\lambda'(t)$, are chosen as $U' = \overline{U}$, $H' = \overline{H}$, $(UH)' = \overline{UH}$, $A' = \overline{A}$, $(A^2)' = \overline{A^2}$, and (by the definition of the subset) $D'(t) = 0$, *where, to be consistent with the general logic of the MCA table,* $\overline{U}$, $\overline{H}$, $\overline{UH}$, $\overline{A}$, *and* $\overline{A^2}$ *pertain to all Indian women who had a fifth birth, not just the subset who did not experience a child death before having a sixth birth or being censored*. A problem arises in that the baseline hazard function that is printed out for (9.3) uses means calculated for the subset. This printed baseline is converted to the new baseline, $\lambda'(t)$, using the approach discussed earlier in Chapter 8, in the context of equation (8.16). If we denote mean values for the subset by a tilde ($\sim$), the conversion formula, based on (8.16), is

$$\lambda'(t) = \tilde{\lambda}(t)e^{b(\overline{U}-\widetilde{U})+c(\overline{H}-\widetilde{H})+d(\overline{UH}-\widetilde{UH})+f(\overline{A}-\widetilde{A})+g(\overline{A^2}-\widetilde{A^2})} \tag{9.4}$$

where in this case b, c, d, f, and g are from the truncated model in (9.3), fitted to the subset of Indian women who had a fifth birth but no child deaths before having a sixth birth or being censored.

We then obtain $\lambda(t)$ as

$$\boxed{\lambda(t) = \lambda'(t)e^{hD(t)}} \tag{9.5}$$

where h is from the full model in equation (9.1), fitted to all Indian women who had a fifth birth. This equation says that $\lambda(t) = \lambda'(t)$ if $D(t) = 0$, and $\lambda(t) = \lambda'(t)e^h$ if $D(t) = 1$. In each of these two cases, as already mentioned, all predictor variables except $D(t)$ are set at their mean values for all Indian women who had a fifth birth, thus conforming to the usual logic of an MCA table.

To obtain a formula for $S(t)$, for use in calculating the child death panel of Table 9.2, we begin with

$$S(t) = e^{-\int_0^t \lambda(x)\,dx} \qquad (7.27 \text{ repeated})$$

which can also be written

$$S(t) = \exp[-\Lambda(t)] \qquad (9.6)$$

where $\Lambda(t)$ is the cumulative hazard. For one-month intervals, $\Lambda(t)$ can be written in discrete form as

$$\Lambda(t) = \sum_{x=0}^{t-1} \lambda_x \qquad (9.7)$$

where λ_t is the hazard rate for the interval t to $t+1$.[1]

Equation (9.5) can also be written in discrete form as

$$\lambda_t = \lambda'_t e^{hD_t} \qquad (9.8)$$

where λ'_t is the baseline hazard from equation (9.5) written in discrete form.

By substituting (9.8) into (9.7), we can rewrite (9.7) as

$$\Lambda(t) = \sum_{x=0}^{t-1} \lambda'_x e^{hD_x} \qquad (9.9)$$

Substituting (9.9) into (9.6), we obtain

$$\boxed{S(t) = \exp\left[-\sum_{x=0}^{t-1} \lambda'_x e^{hD_x}\right]} \qquad (9.10)$$

Let us now denote by t^* the time of a woman's first child death if she had any child deaths. Suppose that she had a child death at $t^* = 24$ months after having a fifth birth. Because of the 12-month time lag, the hazard of a sixth birth does not get multiplied by e^h until 12 months later, at $t = 36$ months. Therefore $D_t = 0$ for $t < 36$, and $D_t = 1$ for $t \geq 36$. Estimated survivorship at 120 months is therefore calculated from (9.10) as

$$\begin{aligned} S(120) &= \exp[-\Lambda(120)] \\ &= \exp\left[-\left(\sum_{t=0}^{35} \lambda'_t + \sum_{t=36}^{119} \lambda'_t e^h\right)\right], \quad t^* = 24 \qquad (9.11) \end{aligned}$$

[1]Computer programs for hazard models print out $\Lambda(t)$ at one-unit time intervals, which in our examples are one-month intervals.

The corresponding value of $F(120)$ is then calculated as $F(120) = 1 - S(120)$. The result of this calculation is .82, as shown in the child death panel of Table 9.2. The other entries in this panel are similarly calculated.

The strategy underlying the construction of the MCA table in Table 9.2 may be summarized as follows: In each panel, we try our best to hold all predictor variables constant at their means (but this cannot be done for time-dependent variables, which are left uncontrolled), except for the variable whose effect is being examined in that panel.

9.2. TIME-DEPENDENT COEFFICIENTS

9.2.1. Coefficients and Effects

The hazard model can handle time-dependent coefficients as well as time-dependent predictor variables.

To illustrate time-dependent coefficients, we shall add ethnicity, I (1 if Indian, 0 if Fijian), to the model considered in Section 9.1.3. Recall that the model in Section 9.1.3 was run only for Indians, and that the reason why the model was run only for Indians was that ethnicity, as a predictor variable, behaves nonproportionally in a way that cannot be modeled adequately by a predictor variable with a fixed coefficient (see Figure 8.3 and related discussion in Chapter 8).

How can we model the nonproportionality of ethnicity? If ethnicity were the only predictor variable in the model, and if ethnicity behaved proportionally, the model could be represented as $\lambda(t) = \lambda_f(t)e^{bI}$, and the shape of the curve in Figure 8.3(b), representing the difference between Indians and Fijians in $\log[-\log S(t)]$, would be a straight line, $Y = k$, where $Y = \log[-\log S_i(t)] - \log[-\log S_f(t)]$, where subscripts i and f denote Indians and indigenous Fijians, respectively. [See equation (8.55) of Chapter 8.] In reality, however, the curve in Figure 8.3b is not a straight line, but resembles instead a simple arc. This suggests that the coefficient might be modeled as a simple quadratic function of time, $m + nt + pt^2$. Ethnicity (I) times its coefficient can then be written

$$
\begin{aligned}
(m + nt + pt^2)I &= mI + nIt + pIt^2 \\
&= mI + nJ(t) + pK(t) \qquad (9.12)
\end{aligned}
$$

where we define

$$J(t) \equiv It$$
$$K(t) \equiv It^2$$

The model then becomes

$$\lambda(t) = \bar{\lambda}(t)e^{a+bU+cH+dUH+fA+gA^2+hD(t)+(m+nt+pt^2)I} \tag{9.13}$$

or

$$\lambda(t) = \bar{\lambda}(t)e^{a+bU+cH+dUH+fA+gA^2+hD(t)+mI+nJ(t)+pK(t)} \tag{9.14}$$

In the estimation process, (9.14) *is used, and $J(t)$ and $K(t)$ are treated as separate time-dependent predictor variables with fixed coefficients. By means of this trick, the computer program proceeds as in Section* 9.1 *on time-dependent predictor variables, so that no new programming is needed to deal with the time-dependent coefficient.*

What is the effect of a one-unit increase in I, holding all other predictor variables and t constant? By the same kind of reasoning employed in Chapter 8 for proportional hazard models, this effect is to multiply the hazard by e raised to the power of its coefficient—that is, by $e^{m+nt+pt^2}$.

Is this effect statistically significant? To answer this question, we must test whether $e^{m+nt+pt^2}$ differs significantly from one or, equivalently, whether $m + nt + pt^2$ differs significantly from zero. For this latter test, we have that

$$\begin{aligned} Z &= \frac{m + nt + pt^2}{\sqrt{\mathrm{Var}(m + nt + pt^2)}} \\ &= \frac{m + nt + pt^2}{\sqrt{\mathrm{Var}\,(m) + t^2\,\mathrm{Var}\,(n) + t^4\,\mathrm{Var}\,(p) + 2t\,\mathrm{Cov}(m, n) + 2t^2\,\mathrm{Cov}(m, p) + 2t^3\,\mathrm{Cov}(n, p)}} \end{aligned} \tag{9.15}$$

Note that t is treated as a constant in these formulae. Note also that Z depends on t, so that it is possible for the effect of ethnicity on the hazard to be significant at some values of t and nonsignificant at others. As usual, the values of the variances and covariances in (9.15) are obtained from the regression coefficient covariance matrix or, in

TABLE 9.3. Coefficients and Multiplicative Effects for the Estimated Model $\lambda(t) = \bar{\lambda}(t)e^{a+bU+cH+dUH+fA+gA^2+hD(t)+(m+nt+pt^2)I}$, Where $\lambda(t)$ Is the Hazard of Having a Sixth Birth: Women Who Have Had at Least Five Births: 1974 Fiji Fertility Survey[a]

Predictor Variable	Coefficient (S.E.)	Effect
Residence × education		
U	−.1635 (.0864)	$e^{-.1635+.0021H}$
H	−.1441 (.0768)	$e^{-.1441+.0021U}$
UH	.0021 (.1249)	—
Age		$e^{.0432+2(-.0021)A}$
A	.0432 (.0678)	—
A^2	−.0021 (.0012)	—
Child death		
$D(t)$	.1785* (.0601)	$e^{.1785} = 1.20$
Ethnicity		$e^{1.3723-.0868t+.0008t^2}$
I	1.3723* (.2394)	—
$J(t)\ (= It)$	−.0868* (.0130)	—
$K(t)\ (= It^2)$	.0008* (.0001)	—

[a]An asterisk after a coefficient indicates that it is significant at the 5 percent level.

the case of the variances, by squaring the standard errors of the coefficients.

Table 9.3 shows estimated coefficients and effects for the model in (9.13) and (9.14). The coefficients of I, $J(t)$, and $K(t)$ are significant at the 5 percent level. The effects of age and child death are rather similar to what they were for Indians by themselves in Table 9.1. However, the effects of residence and education differ from corresponding values in Table 9.1. This latter finding indicates that the effects of residence and education differ between Indians and Fijians, suggesting that the model should be rerun with a three-way interaction between residence, education, and ethnicity. Because a three-way interaction would make the model very complicated, especially because the coefficient of ethnicity is time-dependent, we do not pursue this option.

9.2.2. Survival Functions

In a real research project, we might decide to abandon a combined model at this point and instead run separate models for Indians and

Fijians, because the two groups behave so differently. However, for purposes of illustration, let us pursue the model in (9.13) a little further.

It is interesting to examine, for example, how well the model captures the crossover in observed survival curves, noted earlier in Figure 8.3a. Figure 9.1a shows that the estimated survival curves for Indians and Fijians do indeed cross over. The curves in Figure 8.3a and Figure 9.1a are not exactly comparable, however, because Figure

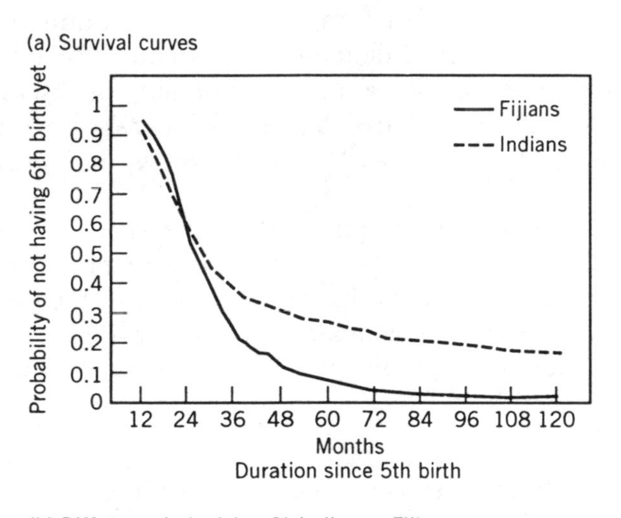

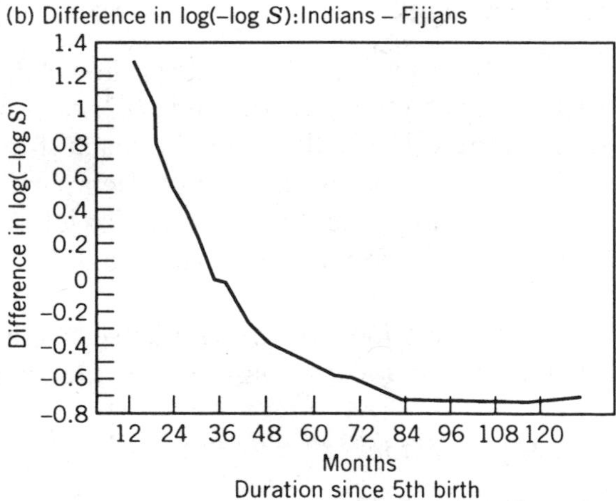

Figure 9.1. Survival curves, $S(t)$, for Indians and Fijians, and differences in $\log[-\log S(t)]$ between Indians and Fijians, based on fitted values of $S(t)$ derived from the model

$$\lambda(t) = \bar{\lambda}(t)e^{a+bU+cH+dUH+fA+gA^2+hD(t)+(m+nt+pt^2)I}$$

8.3a does not control for other predictor variables, whereas Figure 9.1a does.

In Figure 9.1a, the survival curve for Fijians was derived from the proportional hazard model obtained by excluding child death and ethnicity from the model in (9.13) and then fitting the model just to Fijians. This truncated model yielded a discrete hazard function, λ_t^f (where superscript f denotes Fijians), and a survival curve, $S_f(t)$, evaluated at the means of U, H, UH, A, and A^2, where the means are for Fijians and Indians combined, not for Fijians alone. For Indians, the hazard function was then obtained as $\lambda_t^i = \lambda_t^f e^{m+n(t+.5)+p(t+.5)^2}$ (where i denotes Indians and f Fijians; .5 is added to t in order to obtain values at the midpoints of discrete one-year time intervals), the cumulative hazard as $\Lambda_i(t) = \Sigma_{x=0}^{t-1}\lambda_x^i$, and the survival function as $S_i(t) = e^{-\Lambda_i(t)}$, where m, n, and p are from the full model in (9.13).

Figure 9.1b shows the graph of the difference in $\log[-\log S(t)]$ between Indians and Fijians, based on these estimated survival curves. The graph is rather similar to that shown earlier in Figure 8.3a. Again the graphs are not exactly comparable, because Figure 9.1a incorporates controls for other predictor variables whereas Figure 8.3a does not.

9.2.3. MCA Adapted to Hazard Regression with Time-Dependent Coefficients

The strategy for constructing the MCA table, shown in Table 9.4, is similar to that used earlier to construct Table 9.2.

There are four panels in Table 9.4, one each for residence × education, age, child death, and ethnicity. For each panel, the strategy is to choose a baseline as close to the ideal baseline as possible. Coefficients from the full model, given in Table 9.3, are used in formulae in all the panels.

The baseline hazard function for the residence × education panel and the age panel is obtained as the baseline for the truncated model

$$\lambda(t) = \bar{\lambda}(t)e^{a+bU+cH+dUH+fA+gA^2} \tag{9.16}$$

for all women in the sample who had at least five births. The time-dependent terms $hD(t)$ and $mI + nJ(t) + pK(t)$ are excluded. In calculating the MCA table entries for the residence × education and age panels, one uses $\bar{S}(120)$ derived from $\bar{\lambda}(t)$ in the truncated model in (9.16), but coefficients b, c, d, f, and g from the full model

TABLE 9.4. MCA Table of Adjusted Values of $F(120)$ for the Estimated Model $\lambda(t) = \bar{\lambda}(t)^{e^{a + bU + cH + dUH + fA + gA^2 + hD(t) + (m + nt + pt^2)I}}$, Where $\lambda(t)$ is the Hazard of Having a Sixth Birth: Women Who Have Had at Least Five Births, 1974 Fiji Fertility Survey[a]

Predictor Variable	n	Adjusted Value of $F(120)$: Formula	Adjusted Value of $F(120)$: Numerical Value
Residence × education			
Rural, low		$1 - [\bar{S}(120)]^{e^{-b\bar{U} - c\bar{H} - d\overline{UH}}}$	.85
Rural, high		$1 - [\bar{S}(120)]^{e^{-b\bar{U} + c(1-\bar{H}) - d\overline{UH}}}$	.85
Urban, low		$1 - [\bar{S}(120)]^{e^{b(1-\bar{U}) - c\bar{H} - d\overline{UH}}}$	.80
Urban, high		$1 - [\bar{S}(120)]^{e^{b(1-\bar{U}) + c(1-\bar{H}) + d(1-\overline{UH}}}$	.80
Age			
25	—	$1 - [\bar{S}(120)]^{e^{f(25-\bar{A}) + g(625 - \bar{A}2)}}$	.90
30	—	$1 - [\bar{S}(120)]^{e^{f(30-\bar{A}) + g(900 - \bar{A}2)}}$	.80
35	—	$1 - [\bar{S}(120)]^{e^{f(35-\bar{A}) + g(1225 - \bar{A}2)}}$	.63
40	—	$1 - [\bar{S}(120)]^{e^{f(40-\bar{A}) + g(1600 - \bar{A}2)}}$	.43
Child death			
At $t = -12$	—	$1 - \exp\left[-\sum_{t=0}^{119} \lambda'_t e^h\right]$	.87
At $t = 0$	—	$1 - \exp\left[-\left(\sum_{t=0}^{11} \lambda'_t + \sum_{t=12}^{119} \lambda'_t e^h\right)\right]$	.87
At $t = 24$	—	$1 - \exp\left[-\left(\sum_{t=0}^{35} \lambda'_t + \sum_{t=36}^{119} \lambda'_t e^h\right)\right]$	.85
At $t = 48$	—	$1 - \exp\left[-\left(\sum_{t=0}^{59} \lambda'_t + \sum_{t=60}^{119} \lambda'_t e^h\right)\right]$	.84
At $t = 60$	—	$1 - \exp\left[-\left(\sum_{t=0}^{71} \lambda'_t + \sum_{t=72}^{119} \lambda'_t e^h\right)\right]$	.83
No child dies	—	$1 - \exp\left[\sum_{t=0}^{119} \lambda'_t\right]$	.79
Ethnicity			
Indians		$1 - \left[\exp\left(-\sum_{t=0}^{119} \lambda_t^f e^{m + n(t+.5) + p(t+.5)^2}\right)\right]$	.82
Fijians		$1 - \left[\exp\left(-\sum_{t=0}^{119} \lambda_t^f\right)\right]$	.97

[a]See text for explanation of how the baseline hazard function is defined for each of the four panels of this table. In every panel, coefficients appearing in calculation formulae are taken from the full model. See also the footnote to Table 9.2.

in (9.14). Child death and ethnicity are sources of unobserved heterogeneity that bias the baseline hazard but not the coefficients appearing in the calculation formulae, because the coefficients are calculated from the full model in which child death (which is time-varying) and ethnicity (which has a time-varying effect) are controlled.

The baseline hazard function for the child death panel is based on the truncated model in (9.16) for Indian and Fijian women combined who had a fifth birth but no child deaths before having a sixth birth or censoring. One then proceeds in essentially the same way as in the child death panel in Table 9.2. In the child death panel in Table 9.4, ethnicity is a source of unobserved heterogeneity that biases the baseline hazard but not the coefficients appearing in the calculation formulae, because the coefficients are calculated from the full model in (9.14) in which ethnicity is controlled.

The baseline hazard function for the ethnicity panel is based on the truncated model in (9.16) for Fijian women who have had at least five births. The approach is similar to that used in the child death panel. In the ethnicity panel, child death is a source of unobserved heterogeneity that biases the baseline hazard but not the coefficients appearing in the calculation formulae, because the coefficients are calculated from the full model in (9.14) in which child death is controlled.

9.3. FURTHER READING

See Kalbfleisch and Prentice (1980) and Cox and Oakes (1984).

APPENDIX A

SAMPLE COMPUTER PROGRAMS

This Appendix of selected computer programs is intended to facilitate the work of those who apply the statistical methods described in the text of this book. The SAS programs (SAS Institute, 1990) and BMDP programs (Dixon, 1992) listed are the programs that were actually used to estimate models and generate tables in the text. These programs were run on an IBM3081 mainframe computer with operating system MVS/SP-JES2 1.3.6. LIMDEP programs (Green, 1992) are also provided, especially for those who intend to use personal computers. LIMDEP can be used for all the statistical methods described in this book.

The sample computer programs use data from the 1974 Fiji Fertility Survey, which was part of the World Fertility Survey program. The Fiji Fertility Survey computer tape is in the public domain and has provided a convenient source of constructed examples that appear in the text. Selected variables were extracted and saved in a file called FIJIDATA, which is described below. The SAS and LIMDEP programs assume the existence of the FIJIDATA file.

A.1. DESCRIPTION OF THE FIJIDATA FILE

Variable	Description and Codes
DINTV	Date of interview in century-month code (months are numbered consecutively from January 1990, for which the code is 1)

DBIRTH	Date of birth in century-month code
AGE	Age in completed years
AGEM	Age at first marriage in completed years
CEB	Number of children ever born
CMETHOD	Contraceptive method currently used
	0 None
	1 Pill
	2 IUD
	3 Other female scientific method
	4 Douche
	5 Condom
	6 Rhythm method
	7 Withdrawal
	8 Abstinence
	9 Female sterilization
	10 Male sterilization
	11 Injection
	15 Other methods
	88 Not exposed
	99 Not stated
RES	Type of place of residence
	1 Suva
	2 Other urban
	3 Rural
	99 Not stated
ED	Number of years of education
ETHNIC	Ethnic group
	1 Fijian
	2 Indian
	3 European
	4 Part European
	5 Pacific Islander
	6 Rotuman
	7 Chinese
	8 Other
DBIRTH1	Date of birth of 1st child coded as century-month
AGEDYR1	Age at death of 1st child in completed years
AGEDMON1	Age at death of 1st child in completed months
DBIRTH2	Date of birth of 2nd child, century-month
AGEDYR2	Age at death of 2nd child, years
AGEDMON2	Age at death of 2nd child, months
DBIRTH3	Date of birth of 3rd child, century-month

AGEDYR3	Age at death of 3rd child, years
AGEDMON3	Age at death of 3rd child, months
DBIRTH4	Date of birth of 4th child, century-month
AGEDYR4	Age at death of 4th child, years
AGEDMON4	Age at death of 4th child, months
DBIRTH5	Date of birth of 5th child, century-month
AGEYR5	Age at death of 5th child, years
AGEDMON5	Age at death of 5th child, months
DBIRTH6	Date of birth of 6th child, century-month
AGEDYR6	Age at death of 6th child, years
AGEDMON6	Age at death of 6th child, months

A.2. SAS MAINFRAME PROGRAMS

PROGRAM 1. SAS Program for Multiple Regression (Program Used for Table 2.2)

```
DATA;
SET FIJIDATA;
/*                                                        */
/* Exclude women whose residence is unknown.              */
/* Select only Fijians and Indians aged 35 or older.      */
/*                                                        */
IF ((RES LT 99) AND (ETHNIC LE 2) AND (AGE GE 35));
/*                                                        */
/* Create dummy variables URBAN and INDIAN.               */
/*                                                        */
URBAN=0;
INDIAN=0;
IF (RES LE 2) THEN URBAN=1;
IF (ETHNIC EQ 2) THEN INDIAN=1;
/*                                                        */
/* Create interaction variable UI as URBAN x INDIAN.      */
/*                                                        */
UI=URBAN*INDIAN;
/*                                                        */
/* Create quadratic term AGESQ (age squared).             */
/*                                                        */
AGESQ=AGE*AGE;
/*                                                        */
/* Define the regression model.                           */
/* Option STB gives standardized coefficients.            */
/*                                                        */
PROC REG;
     MODEL CEB=AGE AGESQ ED URBAN INDIAN UI / STB;
```

PROGRAM 2. SAS Program for Partial Correlation Coefficient (Program Used for Table 3.4, Adjusted R Column)

```
DATA ALL;
SET FIJIDATA;
IF ((RES LT 99) AND (ETHNIC LE 2) AND (AGE GE 35));
/*                                                         */
/* Create dummy variables URBAN, INDIAN, MEDIUM, HIGH.     */
/*                                                         */
URBAN=0;
INDIAN=0;
MEDIUM=0;
HIGH=0;
IF (RES LE 2) THEN URBAN=1;
IF (ETHNIC EQ 2) THEN INDIAN=1;
IF ((ED GE 5) AND (ED LE 7)) THEN MEDIUM=1;
IF (ED GE 8) THEN HIGH=1;
/*                                                         */
/* Compute partial correlation coefficient between         */
/* fertility (CEB) and residence (URBAN)  controlling for */
/* education (HIGH, MEDIUM) and ethnicity (INDIAN).        */
/*                                                         */
/* First, do multiple regression and save residuals,       */
/* using original data set named ALL.                      */
/*                                                         */
/* Model statement defines two multiple regressions, one   */
/* with CEB as dependent variable and one with URBAN as    */
/* dependent variable.                                     */
/*                                                         */
/* OUTPUT statement saves two residuals, RFHMI from the    */
/* first regression (CEB) and RUHMI from the second        */
/* regression (URBAN) in data set OUTA.                    */
/*                                                         */
PROC REG DATA=ALL;
MODEL CEB URBAN = HIGH MEDIUM INDIAN / STB;
OUTPUT OUT=OUTA  R=RFHMI RUHMI;
/*                                                         */
/* Next, regress RFHMI on RUHMI using the residuals saved */
/* in data set OUTA.  The resulting R is the partial       */
/* correlation  coefficient.                               */
/*                                                         */
PROC REG DATA=OUTA;
MODEL RFHMI = RUHMI / STB;
/*                                                         */
/* Similarly, compute the partial correlation coefficient */
/* between fertility (CEB) and education (HIGH, MEDIUM)    */
/* controlling for residence and ethnicity.                */
/*                                                         */
PROC REG DATA=ALL;
MODEL CEB HIGH MEDIUM = URBAN INDIAN / STB;
OUTPUT OUT=OUTB  R=RFUI RHUI RMUI;
PROC REG DATA=OUTB;
MODEL RFUI = RHUI RMUI / STB;
```

PROGRAM 3. SAS Program for Logit Regression (Program Used for Table 5.1)

Note: The logit regression program LOGIST is supported by the authors of the program, not by SAS Institute, Inc. The authors are

Frank E. Harrell, Jr. and Bercedis Peterson
Clinical Biostatistics
Box 3363 Duke University Medical Center
Durham NC 27710

```
DATA;
SET FIJIDATA;
/*                                                            */
/* Exclude women whose residence is unknown.                  */
/* Select only Fijians and Indians.                           */
/*                                                            */
IF ((RES LT 99) AND (ETHNIC LE 2));
/*                                                            */
/* Exclude women who did not respond to the question on       */
/* contraceptive use (CMETHOD = 99) and women who are not     */
/* exposed to the risk of conception (CMETHOD = 88), and      */
/* select women under age 45.                                 */
/*                                                            */
IF ((CMETHOD LT 88) and (AGE LT 45));
/*                                                            */
/* Create dummy variables, URBAN, INDIAN, MEDIUM, HIGH,       */
/* interaction term UI, and quadratic term AGESQ.             */
/*                                                            */
URBAN=0;
INDIAN=0;
MEDIUM=0;
HIGH=0;
IF (RES LE 2) THEN URBAN=1;
IF (ETHNIC EQ 2) THEN INDIAN=1;
IF ((ED GE 5) AND (ED LE 7)) THEN MEDIUM=1;
IF (ED GE 8) THEN HIGH=1;
UI=URBAN*INDIAN;
AGESQ=AGE*AGE;
/*                                                            */
/* Create the dependent binary variable USE.                  */
/*                                                            */
USE=0;
IF (CMETHOD GT 0) THEN USE=1;
/*                                                            */
/* Define logistic model.  The option PRINTC prints the       */
/* covariances of the estimated coefficients.                 */
/*                                                            */
PROC LOGIST PRINTC;
MODEL USE = AGE AGESQ MEDIUM HIGH URBAN INDIAN UI;
```

PROGRAM 4. SAS Program for Multinomial Logit Regression (Programs Used for Table 6.4)

Note: The multinomial logit regression program MLOGIT is supported by the author of the program, not by SAS Institute, INC. The author is

Salford Systems
(619) 481-8599

```
DATA;
SET FIJIDATA;
IF ((RES LT 99) AND (ETHNIC LE 2));
IF ((CMETHOD LT 88) AND (AGE LT 45));
URBAN=0;
INDIAN=0;
MEDIUM=0;
HIGH=0;
IF (RES LE 2) THEN URBAN=1;
IF (ETHNIC EQ 2) THEN INDIAN=1;
IF ((ED GE 5) AND (ED LE 7)) THEN MEDIUM=1;
IF (ED GE 8) THEN HIGH=1;
UI=URBAN*INDIAN;
AGESQ=AGE*AGE;
/*                                                          */
/* The dependent variable for the SAS MLOGIT procedure      */
/* needs to be coded as 1, 2, ..., K,  where K is the       */
/* number of possible outcomes of the dependent variable.   */
/* Our dependent variable METHOD will be coded as           */
/*        1 for male or female sterilization,               */
/*        2 for other methods,                              */
/*        3 for not using.                                  */
/* The SAS MLOGIT procedure uses the category represented   */
/* by the largest number (in our example, 3: not using )    */
/* as the reference category.                               */
/*                                                          */
METHOD=3;
IF (CMETHOD GT 0) THEN METHOD=2;
IF ((CMETHOD EQ 9) OR (CMETHOD EQ 10)) THEN METHOD=1;
/*                                                          */
/* In the MLOGIT procedure statement, NCAT indicates the    */
/* number of categories of the dependent variable.          */
/* Option COVAR prints the covariances of the estimated     */
/* coefficients.                                            */
/* PARMCARDS4 and four semicolons are required parts of     */
/* the model specification.  The four semicolons must       */
/* start from column 1.                                     */
/*                                                          */
PROC MLOGIT NCAT=3 COVAR;
  PARMCARDS4;
  MODEL METHOD = AGE AGESQ MEDIUM HIGH URBAN INDIAN UI;
;;;;
```

A.3. PREPARING DATA FOR SURVIVAL ANALYSIS

The most flexible computer programs for survival analysis (life tables and hazard models) are BMDP programs 1L and 2L, as well as LIMDEP programs. Because estimation of the hazard model (especially the model with time-dependent covariates) involves many passes through the data, the LIMDEP procedure for survival analysis requires that the data be kept in memory during execution. For this reason, it is economical, in terms of memory requirements, to create variables to be used for survival analysis in a separate program.

The dependent variable in survival analysis is a life table or, equivalently, a hazard function. When we prepare data for survival analysis, we need to provide the basic information required to build a life table or a hazard function.

The required information includes the survival time and the survival status at the survival time (failure or censored). One has information on failure time for some individuals. For them, the survival time is equal to the failure time (duration between the starting event and the terminating event), and the status is failure. For some individuals, one does not know the failure time because censoring occurred before the failure time. The survival time is then equal to the duration between the starting event and the end of observation, and the status is censored. Status is coded as categorical variable. For example, one may use code 0 for censoring and code 1 for failure.

PROGRAM 5. SAS Program to Create Input Data File for BMDP and LIMDEP Programs for Life-Table and Hazard Model Analysis of Progression from Fifth Birth to Sixth Birth

```
DATA;
SET FIJIDATA;
IF ((RES LT 99) AND (ETHNIC LE 2) AND (CEB GT 4));
URBAN=0;
INDIAN=0;
MEDIUM=0;
HIGH=0;
IF (RES LE 2) THEN URBAN=1;
IF (ETHNIC EQ 2) THEN INDIAN=1;
IF ((ED GE 5) AND (ED LE 7)) THEN MEDIUM=1;
IF (ED GE 8) THEN HIGH=1;
/*                                                           */
/* Compute age of mother in century months at time of 5th */
/* birth.                                                    */
/*                                                           */
AGE5B=DBIRTH5-DBIRTH;
/*                                                           */
/* Compute date of death of first child in century month   */
/* code.  Set this code to 9999 if the 1st child did not   */
/* die.                                                      */
/*                                                           */
DDEATH1=9999;
IF (AGEDYR1 LT 88) THEN
DDEATH1=DBIRTH1+AGEDYR1*12+AGEDMON1+0.5;
/*                                                           */
/* Compute dates of death of 2nd through 5th children.     */
/*                                                           */
DDEATH2=9999;
IF (AGEDYR2 LT 88) THEN
      DDEATH2=DBIRTH2+AGEDYR2*12+AGEDMON2+0.5;
DDEATH3=9999;
IF (AGEDYR3 LT 88) THEN
      DDEATH3=DBIRTH3+AGEDYR3*12+AGEDMON3+0.5;
DDEATH4=9999;
IF (AGEDYR4 LT 88) THEN
      DDEATH4=DBIRTH4+AGEDYR4*12+AGEDMON4+0.5;
DDEATH5=9999;
IF (AGEDYR5 LT 88) THEN
      DDEATH5=DBIRTH5+AGEDYR5*12+AGEDMON5+0.5;
/*                                                           */
/* Find the earliest date of death of a child among 1st    */
/* through 5th births.  The date is coded as 9999 if none */
/* of these children died.                                   */
/*                                                           */
SDEATHT=9999;
IF (DDEATH1 LT SDEATHT) THEN SDEATHT=DDEATH1;
IF (DDEATH2 LT SDEATHT) THEN SDEATHT=DDEATH2;
```

PROGRAM 5. *Continued*

```
IF (DDEATH3 LT SDEATHT) THEN SDEATHT=DDEATH3;
IF (DDEATH4 LT SDEATHT) THEN SDEATHT=DDEATH4;
IF (DDEATH5 LT SDEATHT) THEN SDEATHT=DDEATH5;
/*                                                        */
/* Compute 6th birth interval, setting it equal to 999 if */
/* woman did not have 6th birth before the time of survey.*/
/*                                                        */
INT6=999;
IF (CEB GT 5) THEN INT6==DBIRTH6-DBIRTH5;
/*                                                        */
/* Save selected variables in file BRTHINT6.              */
/*                                                        */
FILE BRTHINT6;
PUT (DINTV DBIRTH5 SDEATHT URBAN HIGH MEDIUM INDIAN
     ED AGE AGE5B INT6)
    (3*5. 4*2. 3*3. 4.);
```

A.4. BMDP MAINFRAME PROGRAMS

PROGRAM 6. BMDP1L Program for Actuarial Life Table: Progression from Fifth Birth to Sixth Birth for Fijians and Indians (Program Used for Table 7.1 and the Actuarial Life Tables in Table 7.2)

```
/ PROBLEM
  TITLE IS 'ACTUARIAL LIFE TABLE'.
#
# This program uses data created by Program 5.
# Four variables are read from the input file, and one
# additional variable (PP) will be created.
#
/ INPUT
  VARIABLES ARE 4.
  UNIT IS 10.
  FORMAT IS '(2F5.0,11x,F2.0,9x,F4.0)'.
/ VARIABLES
  NAMES ARE DINTV, DBIRTH5, INDIAN, INT6, PP.
  ADD=1.
/ TRAN
#
# Calculate interval between birth of 5th child and survey.
#
  OPEN=DINTV-DBIRTH5.
#
# The variable PP indicates whether woman had 6th birth or
# not.  For construction of life table describing
# progression to 6th birth, the birth of 6th child is
# considered as failure (PP=1), and women who did not have
# a 6th birth before the survey (indicated by INT6=999) are
# considered as censored (PP=0).
#
  PP=0.
  IF (INT6 LT 999) THEN PP=1.
#
# For censored cases, survival time is the open birth
# interval, i.e., the interval between 5th birth and
# interview.
#
  IF (PP=0) THEN INT6=OPEN.
#
# We want to close the life table at 120 months.  In order
# to do this, we impose censoring at 120 months on those
# who are still surviving at 120 months.
#
  IF (INT6 GT 120) THEN PP=0.
  IF (INT6 GT 120) THEN INT6=120.
/ FORM
#
# Identify INT6 as the survival time, and PP as the
# indicator for censoring/failure with code 0 indicating
```

PROGRAM 6. *Continued*

```
# censoring and code 1 indicating failure.
#
  TIME IS INT6.
  STATUS IS PP.
  RESPONSE IS 1.
/ ESTIMATE
#
# Either an actuarial life table or a product-limit life
# table can be estimated.  The actuarial method is chosen.
#
  METHOD IS LIFE.
#
# Compute life table using one-month intervals.
#
  WIDTH IS 1.
#
# We estimate separate life tables for Indians (INDIAN=1)
# and Fijians (INDIAN=0).
#
  GROUP IS INDIAN.
/ GROUP
#
# Identify codes and labels for the grouping variable.
# The grouping variable INDIAN is the third variable.
#
  CODES(3) ARE       0,       1.
  NAMES(3) ARE FIJIANS, INDIANS.
/ END
```

PROGRAM 7. BMDP1L Program for Product-Limit Life Table: Progression from Fifth Birth to Sixth Birth for Fijians and Indians (Programs Used for Product-Limit Life Tables in Table 7.2)

Note: This program is exactly same as Program 6 except that, in the ESTIMATE paragraph, the product-limit life table is chosen, and a statistical test of whether two life tables differ significantly is done.

```
/ PROBLEM
  TITLE IS 'PRODUCT-LIMIT LIFE TABLE'.
/ INPUT
  VARIABLES ARE 4.
  UNIT IS 10.
  FORMAT IS '(2F5.0,11x,F2.0,9x,F4.0)'.
/ VARIABLES
  NAMES ARE DINTV, DBIRTH5, INDIAN, INT6, PP.
  ADD=1.
/ TRAN
  OPEN=DINTV-DBIRTH5.
  PP=0.
  IF (INT6 LT 999) THEN PP=1.
  IF (PP=0) THEN INT6=OPEN.
  IF (INT6 GT 120) THEN PP=0.
  IF (INT6 GT 120) THEN INT6=120.
/ FORM
  TIME IS INT6.
  STATUS IS PP.
  RESPONSE IS 1.
/ ESTIMATE
#
# Choose product-limit life table.
#
  METHOD IS PRODUCT.
  GROUP IS INDIAN.
#
# Test if two life tables differ significantly from each
# other. Choose both Breslow test and Mantel test.
#
  STATISTICS ARE BRESLOW, MANTEL.
/ GROUP
  CODES(3) ARE       0,       1.
  NAMES(3) ARE FIJIANS, INDIANS.
/ END
```

PROGRAM 8. BMDP2L Program for Proportional Hazard Model: Progression from Fifth to Sixth Birth for Indians (Program Used for Table 8.1)

```
/ PROBLEM
  TITLE IS 'PROPORTIONAL HAZARD MODEL, INDIANS'.
#
# This program uses data created by Program 5.
# Ten variables are read from the input file, and four
# additional variables (PP, MH, UH, and AGE5BSQ) will be
# created.
#
/ INPUT
  VARIABLES ARE 10.
  UNIT IS 10.
  FORMAT IS '(2F5.0,5x,4F2.0,3F3.0,F4.0)'.
/ VARIABLES
  NAMES ARE DINTV, DBIRTH5, URBAN, HIGH, MEDIUM, INDIAN, ED,
  AGE, AGE5B, INT6, PP, MH, UH, AGE5BSQ.
  ADD=4.
/ TRAN
#
# First, select Indians for analysis
#
  USE=(INDIAN EQ 1).
#
# Create new variables
#
  OPEN=DINTV-DBIRTH5.
  PP=0.
  IF (INT6 LT 999) THEN PP=1.
  IF (PP=0) THEN INT6=OPEN.
  IF (INT6 GT 120) THEN PP=0.
  IF (INT6 GT 120) THEN INT6=120.
#
# MH is the new dummy variable indicating high level of
# education.
#
  MH=0.
  IF ((MEDIUM  EQ 1) OR (HIGH EQ 1)) THEN MH=1.
#
# UH is the interaction term between urban residence
# and high level of education
#
  UH=URBAN*MH.
#
# AGE5B is mother's age at the time of 5th birth.
# AGE5BSQ is AGE5B x AGE5B.
#
  AGE5B=AGE5B/12.
  AGE5BSQ=AGE5B*AGE5B.
/ FORM
```

PROGRAM 8. *Continued*

```
  TIME IS INT6.
  STATUS IS PP.
  RESPONSE IS 1.
/ REGRESSION
#
# Define the hazard model to be estimated.
#
  COVARIATES ARE URBAN, MH, UH, AGE5B, AGE5BSQ.
/ PRINT
#
# Print covariances of the estimated coefficients and
# baseline survival function.
#
  COVARIANCE.
  SURVIVAL.
/ END
```

PROGRAM 9. BMDP2L Program for Hazard Model with Time-Dependent Covariate (Program Used for Table 9.2)

```
/ PROBLEM
  TITLE IS 'HAZARD MODEL WITH TIME DEPENDENCE, INDIANS'.
#
# This program uses data created by Program 5.
# Ten variables are read from the input file, and four
# additional variables (PP, MH, UH, and AGE5BSQ) will be
# created.
#
/ INPUT
  VARIABLES ARE 11.
  UNIT IS 10.
  FORMAT IS '(3F5.0,4F2.0,3F3.0,F4.0)'.
/ VARIABLES
  NAMES ARE DINTV, DBIRTH5, SDEATHT,
  URBAN, HIGH, MEDIUM, INDIAN, ED, AGE, AGE5B, INT6,
  PP, MH, UH, AGE5BSQ.
  ADD=4.
/ TRAN
#
# Select Indians for analysis.
#
  USE=(INDIAN EQ 1).
  OPEN=DINTV-DBIRTH5.
  PP=0.
  IF (INT6 LT 999) THEN PP=1.
  IF (PP=0) THEN INT6=OPEN.
  IF (INT6 GT 120) THEN PP=0.
  IF (INT6 GT 120) THEN INT6=120.
  MH=0.
  IF ((MEDIUM  EQ 1) OR (HIGH EQ 1)) THEN MH=1.
  UH=URBAN*MH.
  AGE5B=AGE5B/12.
  AGE5BSQ=AGE5B*AGE5B.
/ FORM
  TIME IS INT6.
  STATUS IS PP.
  RESPONSE IS 1.
/ REGRESSION
  COVARIATES ARE URBAN, MH, UH, AGE5B, AGE5BSQ.
#
# The hazard model has a time-dependent covariate SIBDEATH,
# which will be defined in the FUNCTION paragraph.  The
# FUNCTION paragraph will use the variable SDEATHT.  See
# Program 5 for definition of SDEATHT.
#
  ADD=SIBDEATH.
  AUXILIARY=SDEATHT.
/ FUNCTION
```

PROGRAM 9. *Continued*

```
  SIBDEATH=0.
#
# SIBDEATH is a function of time.  Its value is 0 if there
# was no death among the first 5 children as of 11 months
# ago.  The value changes to one 12 months after death of
# any child among the first five children.  The value stays
# as 1 thereafter.
#
  SDEATH12=SDEATHT+12.
  IF (TIME GT SDEATH12) THEN SIBDEATH=1.
/ PRINT
#
# Print the covariances of the estimated coefficients.
#
  COVARIANCE.
/ END
```

PROGRAM 10. BMDP2L Program for Hazard Model with Time-Dependent Covariate and Time-Dependent Effect (Program Used for Table 9.3)

```
/ PROBLEM
  TITLE IS 'HAZARD MODEL WITH TIME DEPENDENCE, INDIANS'.
#
# This program uses data created by Program 5.
# Ten variables are read from the input file, and four
# additional variables (PP, MH, UH, and AGE5BSQ) will be
# created.
#
/ INPUT
  VARIABLES ARE 11.
  UNIT IS 10.
  FORMAT IS '(3F5.0,4F2.0,3F3.0,F4.0)'.
/ VARIABLES
  NAMES ARE DINTV, DBIRTH5, SDEATHT,
  URBAN, HIGH, MEDIUM, INDIAN, ED, AGE, AGE5B, INT6,
  PP, MH, UH, AGE5BSQ.
  ADD=4.
/ TRAN
  USE=(INDIAN EQ 1).
  OPEN=DINTV-DBIRTH5.
  PP=0.
  IF (INT6 LT 999) THEN PP=1.
  IF (PP=0) THEN INT6=OPEN.
  IF (INT6 GT 120) THEN PP=0.
  IF (INT6 GT 120) THEN INT6=120.
  MH=0.
  IF ((MEDIUM  EQ 1) OR (HIGH EQ 1)) THEN MH=1.
  UH=URBAN*MH.
  AGE5B=AGE5B/12.
  AGE5BSQ=AGE5B*AGE5B.
/ FORM
  TIME IS INT6.
  STATUS IS PP.
  RESPONSE IS 1.
/ REGRESSION
  COVARIATES ARE URBAN, MH, UH, AGE5B, AGE5BSQ, INDIAN.
#
# Three time dependent covariates, SIBDEATH, INDIANT, and
# INDIANT2 will be defined and included in the hazard model.
#
  ADD=SIBDEATH, INDIANT, INDIANT2.
#
# Variables SDEATHT and INDIAN will be used to define time-
# dependent covariates.
#
  AUXILIARY=SDEATHT, INDIAN.
/ FUNCTION
#
```

PROGRAM 10. *Continued*

```
# SIBDEATH is defined in the same way as in Program 9.
#
  SIBDEATH=0.
  SDEATH12=SDEATHT+12.
  IF (TIME GT SDEATH12) THEN SIBDEATH=1.
#
# INDIANT is defined as INDIAN x (t).
# INDIANT2 is defined as INDIAN x (t x t).
#
  INDIANT=INDIAN*TIME.
  INDIANT2=INDIAN*TIME*TIME.
/ PRINT
  COVARIANCE.
/ END
```

A.5. LIMDEP PROGRAMS FOR IBM-COMPATIBLE PERSONAL COMPUTERS

PROGRAM 11. LIMDEP Program for Multiple Regression (Program Used for Table 2.2)

```
?
? Exclude women whose residence is unknown.
? Select only Fijians and Indians aged 35 or older.
?
REJECT; RES=99$
REJECT; ETHNIC>2$
REJECT; AGE<35$
?
? Create dummy variables URBAN and INDIAN.
?
CREATE; IF (RES <3) URBAN=1; (ELSE) URBAN=0$
CREATE; IF (ETHNIC = 2) INDIAN=1; (ELSE) INDIAN=0$
?
? Create the interaction variable UI as URBAN x INDIAN.
?
CREATE; UI=URBAN*INDIAN$
?
? Create quadratic term AGESQ (age squared).
?
CREATE; AGESQ=AGE*AGE$
?
? Define the regression model.
? It is important to remember to include ONE as one of the
? independent variables.  The model will not have a constant
? term (intercept) if it is not included.
?
CRMODEL; LHS=CEB;
RHS=ONE, AGE, AGESQ, ED, URBAN, INDIAN, UI$
```

PROGRAM 12. LIMDEP Program for Partial Correlation Coefficient (Program Used for Table 3.4, Adjusted R Column)

```
This program computes two partial correlation coefficients,
one between fertility (CEB) and residence (URBAN),
controlling for education (HIGH and MEDIUM) and ethnicity
(INDIAN); and one between fertility (CEB) and education
(MEDIUM and HIGH), controlling for residence (URBAN) and
ethnicity (INDIAN).

?
? Exclude women whose residence is unknown.
? Select only Fijians and Indians aged 35 or older.
?
REJECT; RES=99$
REJECT; ETHNIC>2$
REJECT; AGE<35$
?
? Create dummy variables URBAN, INDIAN, MEIDUM, and HIGH.
?
CREATE; IF (RES <3) URBAN=1; (ELSE) URBAN=0$
CREATE; IF (ETHNIC = 2) INDIAN=1; (ELSE) INDIAN=0$
CREATE; IF (ED > 4) MEDIUM=1$
CREATE; IF (ED > 7) HIGH=1; (ELSE) HIGH=0$
CREATE; IF (HIGH = 1) MEDIUM=0$
?
? Regression of fertility (CEB) on education (HIGH, MEDIUM)
? and ethnicity (INDIAN).  Save residual as CRES1.
?
CRMODEL; LHS=CEB; RHS=ONE, HIGH, MEDIUM, INDIAN; RES=CRES1$
?
? Regression of residence (URBAN) on education (HIGH,
? MEDIUM) and ethnicity (INDIAN).  Save residual as URES.
?
CRMODEL; LHS=URBAN; RHS=ONE, HIGH, MEDIUM, INDIAN; RES=URES$
?
? Regression of CRES1 on URES.  The resulting R is the
? partial correlation coefficient.
?
CRMODEL; LHS=CRES1; RHS=ONE, URES$
?
? Compute the multiple correlation coefficient between
? fertility (CEB) and education (MEDIUM, HIGH) controlling
? for residence (URBAN) and ethnicity (INDIAN).
?
CRMODEL; LHS=CEB; RHS=ONE, URBAN, INDIAN; RES=CRES2$
CRMODEL; LHS=MEDIUM; RHS=ONE, URBAN, INDIAN; RES=MRES$
CRMODEL; LHS=HIGH; RHS=ONE, URBAN, INDIAN; RES=HRES$
CRMODEL; LHS=CRES2; RHS=ONE, MRES, HRES$
```

PROGRAM 13. LIMDEP Program for Logit Regression (Program Used for Table 5.1)

```
?
? Exclude women whose residence is unknown.
? Select only Fijians and Indians aged 35 or older.
?
REJECT; RES=99$
REJECT; ETHNIC>2$
REJECT; AGE>44$
?
? Create dummy variables URBAN, INDIAN, MEIDUM, and HIGH.
?
CREATE; IF (RES <3) URBAN=1; (ELSE) URBAN=0$
CREATE; IF (ETHNIC = 2) INDIAN=1; (ELSE) INDIAN=0$
CREATE; IF (ED > 4) MEDIUM=1; (ELSE) MEDIUM=0$
CREATE; IF (ED > 7) HIGH=1; (ELSE) HIGH=0$
CREATE; IF (HIGH = 1) MEDIUM=0$
?
? Create the interaction variable UI as URBAN x INDIAN.
?
CREATE; UI=URBAN*INDIAN$
?
? Create quadratic term AGESQ (age squared).
?
CREATE; AGESQ=AGE*AGE$
?
? Define the dependent variable of the logistic regression.
?
CREATE; IF (CMETHOD > 0) USE=1; (ELSE) USE=0$
?
? Define the logistic regression model.
? Do not forget to inlcude ONE on the right hand side for
? the constant term.
?
LOGIT; LHS=USE;
RHS=ONE, AGE, AGESQ, MEDIUM, HIGH, URBAN, INDIAN, UI$
```

PROGRAM 14. LIMDEP Program for Multinomial Logit Regression (Program Used for Table 6.4)

```
?
? Exclude women whose residence is unknown.
? Select only Fijians and Indians aged 35 and older.
?
REJECT; RES=99$
REJECT; ETHNIC>2$
REJECT; AGE>44$
?
? Create dummy variables URBAN, INDIAN, MEDIUM, and HIGH.
?
CREATE; IF (RES <3) URBAN=1; (ELSE) URBAN=0$
CREATE; IF (ETHNIC = 2) INDIAN=1; (ELSE) INDIAN=0$
CREATE; IF (ED > 4) MEDIUM=1; (ELSE) MEDIUM=0$
CREATE; IF (ED > 7) HIGH=1; (ELSE) HIGH=0$
CREATE; IF (HIGH = 1) MEDIUM=0$
?
? Create the interaction term, and the quadratic term.
?
CREATE; UI=URBAN*INDIAN$
CREATE; AGESQ=AGE*AGE$
?
? The dependent variable METHOD is coded as 0 if the woman
? is not using any method, 1 if she is using some method and
? neither she nor her husband is sterilized, and 2 if either
? she or her husband is sterilized.  The LIMDEP multinomial
? logistic regresison program LOGIST (same as binary) uses
? category 0 (not using any method) as the reference
? category of the dependent variable.
?
CREATE; IF (CMETHOD = 9) FST=1; (ELSE) FST=0$
CREATE; IF (CMETHOD = 10) MST=1; (ELSE) MST=0$
CREATE; IF (CMETHOD = 0) USE=0; (ELSE) USE=1$
CREATE; TMETHOD=FST+MST+USE$
CREATE; IF (TMETHOD > 2) METHOD=2; (ELSE) METHOD=TMETHOD$
?
? Define the multinomial logistic regression model.
? Do not forget to include ONE on the right-hand side for
? the constant term.
?
LOGIT; LHS=METHOD;
RHS=ONE, AGE, AGESQ, MEDIUM, HIGH, URBAN, INDIAN, UI$
```

PROGRAM 15. LIMDEP Program for Life Table (Actuarial and Product-Limit): Progression from Fifth Birth to Sixth Birth, Indians (Program Used for Table 7.2)

Note: This program used the data file created by Program 5.

```
REJECT; INDIAN=0$
?
? Calculate interval between birth of 5th child and survey.
?
CREATE; OPEN=DINTV-DBIRTH5$
?
? The variable PP indicates whether woman had 6th birth or
? not.  For construction of life table describing
? progression to 6th birth, the birth of 6th child is
? considered as failure (PP=1), and women who did not have
? the 6th birth before the survey (indicated by INT6=999)
? are considered as censored (PP=0).
?
CREATE; IF (INT6 = 999) PP=0; (ELSE) PP=1$
?
? For censored cases, survival time is the open birth
? interval, i.e., the interval between 5th birth and
? survey.
?
CREATE; IF (INT6 = 999) INT6=OPEN$
?
? We want to close the life table at 120 months.  In order
? to do this, we impose censoring at 120 months on those
? who are still surviving at 120 months.
?
CREATE; IF (INT6>120) PP=0$
CREATE; IF (INT6>120) INT6=120$
REJECT; INT6<0$
?
? The survival procedure without RHS (right-hand side)
? computes actuarial life tables. The LHS (left-hand side)
? of the survival procedure takes two variables, one for
? survival time, and one for censoring/failure information.
? Failure is indicated by code 1, and censoring is indicated
? by code 0.
? LIST requests listing of survival probabilities estimated
? by product-limit method.  If LIST is not included, the
? program produces only actuarial life table.
? The number of intervals (equally spaced) to be used for
? actuarial life table is indicated by INT.
?
SURVIVAL; LHS=INT6, PP; LIST; INT=120$
```

PROGRAM 16. LIMDEP Program for Proportional Hazard Model: Progression from Fifth Birth to Sixth Birth, Indians (Program Used for Table 8.1)

```
?
? See comments in Program 15 for definitions of created
? variables.
?
REJECT; INDIAN=0$
CREATE; OPEN=DINTV-DBIRTH5$
CREATE; IF (INT6 = 999) pp=0; (ELSE) PP=1$
CREATE; IF (INT6 = 999) INT6=OPEN$
CREATE; IF (INT6>120) PP=0$
CREATE; IF (INT6>120) INT6=120$
CREATE; IF (ED > 4) MH=1; (ELSE) MH=0$
CREATE; UH=URBAN*MH$
CREATE; AGE5B=AGE5B/12$
CREATE; AGE5BSQ=AGE5B*AGE5B$
REJECT; INT6<0$
?
? The procedure SURVIVAL with RHS (covariates) estimates the
? hazard model.
?
SURVIVAL; LHS=INT6, PP; RHS=URBAN, MH, UH, AGE5B, AGE5BSQ$
```

Covariate: Progression from Fifth Birth to Sixth Birth, Indians (Program Used for Table 9.2)

```
?
? See comments in Program 15 for definitions of created
? variables.
?
REJECT; INDIAN=0$
CREATE; OPEN=DINTV-DBIRTH5$
CREATE; IF (INT6 = 999) pp=0; (ELSE) PP=1$
CREATE; IF (INT6 = 999) INT6=OPEN$
CREATE; IF (INT6>120) PP=0$
CREATE; IF (INT6>120) INT6=120$
CREATE; IF (ED > 4) MH=1; (ELSE) MH=0$
CREATE; UH=URBAN*MH$
CREATE; AGE5B=AGE5B/12$
CREATE; AGE5BSQ=AGE5B*AGE5B$
CREATE; SDEATH12=SDEATHT+12$
REJECT; INT6<0$
?
? The time-varying covariate TVC is 1 if (INT6)>SDEATH12 and
? 0 otherwise.
?
SURVIVAL; LHS=INT6, PP; RHS=URBAN, MH, UH, AGE5B, AGE5BSQ;
TVC=(INT6)&SDEATH12$
```

PROGRAM 18. LIMDEP Program for Hazard Model with Time-Dependent Covariate and Time-Dependent Effect: Progression from Fifth Birth to Sixth Birth, Indians (Program Used for Table 9.3)

```
?
? See comments in Program 15 for definitions of created
? variables.
?
REJECT; INDIAN=0$
CREATE; OPEN=DINTV-DBIRTH5$
CREATE; IF (INT6 = 999) pp=0; (ELSE) PP=1$
CREATE; IF (INT6 = 999) INT6=OPEN$
CREATE; IF (INT6>120) PP=0$
CREATE; IF (INT6>120) INT6=120$
CREATE; IF (ED > 4) MH=1; (ELSE) MH=0$
CREATE; UH=URBAN*MH$
CREATE; AGE5B=AGE5B/12$
CREATE; AGE5BSQ=AGE5B*AGE5B$
CREATE; SDEATH12=SDEATHT+12$
REJECT; INT6<0$
?
? The SURVIVAL model has three time-dependent covariates
? defined by three TVC statements.
? The first time-varying covariate TVC is 1 if
? (INT6)>SDEATH12 and 0 otherwise.
? The second and third time-dependent covariates are self-
? explanatory.
?
SURVIVAL; LHS=INT6, PP; RHS=URBAN, MH, UH, AGE5B, AGE5BSQ;
TVC=(INT6)&SDEATH12;
TVC=(INT6)*INDIAN;
TVC=(INT6)*INDIAN*INDIAN$
```

APPENDIX B

STATISTICAL REFERENCE TABLES

TABLE B.1. Student's t Critical Points

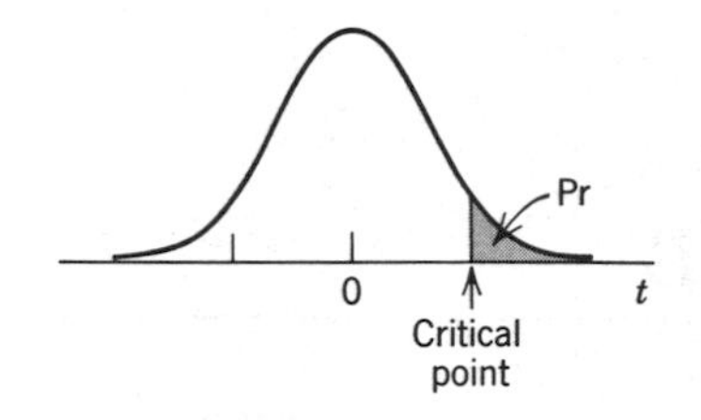

Pr / *d.f.*	*.25*	*.10*	*.05*	*.025*	*.010*	*.005*	*.0025*	*.0010*	*.0005*
1	1.000	3.078	6.314	12.706	31.821	63.637	127.32	318.31	636.62
2	.816	1.886	2.920	4.303	6.965	9.925	14.089	22.326	31.598
3	.765	1.638	2.353	3.182	4.541	5.841	7.453	10.213	12.924
4	.741	1.533	2.132	2.776	3.747	4.604	5.598	7.173	8.610
5	.727	1.476	2.015	2.571	3.365	4.032	4.773	5.893	6.869
6	.718	1.440	1.943	2.447	3.143	3.707	4.317	5.208	5.959
7	.711	1.415	1.895	2.365	2.998	3.499	4.020	4.785	5.408
8	.706	1.397	1.860	2.306	2.896	3.355	3.833	4.501	5.041
9	.703	1.383	1.833	2.262	2.821	3.250	3.690	4.297	4.781
10	.700	1.372	1.812	2.228	2.764	3.169	3.581	4.144	4.537
11	.697	1.363	1.796	2.201	2.718	3.106	3.497	4.025	4.437
12	.695	1.356	1.782	2.179	2.681	3.055	3.428	3.930	4.318
13	.694	1.350	1.771	2.160	2.650	3.012	3.372	3.852	4.221
14	.692	1.345	1.761	2.145	2.624	2.977	3.326	3.787	4.140
15	.691	1.341	1.753	2.131	2.602	2.947	3.286	3.733	4.073
16	.690	1.337	1.746	2.120	2.583	2.921	3.252	3.686	4.015
17	.689	1.333	1.740	2.110	2.567	2.898	3.222	3.646	3.965
18	.688	1.330	1.734	2.101	2.552	2.878	3.197	3.610	3.922
19	.688	1.328	1.729	2.093	2.539	2.861	3.174	3.579	3.883
20	.687	1.325	1.725	2.086	2.528	2.845	3.153	3.552	3.850
21	.686	1.323	1.721	2.080	2.518	2.831	3.135	3.257	3.189
22	.686	1.321	1.717	2.074	2.508	2.819	3.119	3.505	3.792
23	.685	1.319	1.714	2.069	2.500	2.807	3.104	3.485	3.767
24	.685	1.318	1.711	2.064	2.492	2.797	3.091	3.467	3.745
25	.684	1.316	1.708	2.060	2.485	2.787	3.078	3.450	3.725
26	.684	1.315	1.706	2.056	2.479	2.779	3.067	3.435	3.707
27	.684	1.314	1.703	2.052	2.473	2.771	3.057	3.421	3.690
28	.683	1.313	1.701	2.048	2.467	2.763	3.047	3.408	3.674
29	.683	1.311	1.699	2.045	2.462	2.756	3.038	3.396	3.659
30	.683	1.310	1.697	2.042	2.457	2.750	3.030	3.385	3.646
40	.681	1.303	1.684	2.021	2.423	2.704	2.971	3.307	3.551
60	.679	1.296	1.671	2.000	2.390	2.660	2.915	3.232	3.460
120	.677	1.289	1.658	1.980	2.358	2.617	2.860	3.160	3.373
∞	.674	1.282	1.645	1.960	2.326	2.576	2.807	3.090	[illegible]

Source: Wonnacott and Wonnacott, *Econometrics*, (1970), p. 539. Copyright © 1979 John Wiley and Sons, Inc. Reprinted by permission.

TABLE B.2. Standard Normal Cumulative Probability in Right-Hand Tail (For Negative Values of z, Areas Are Found by Symmetry)

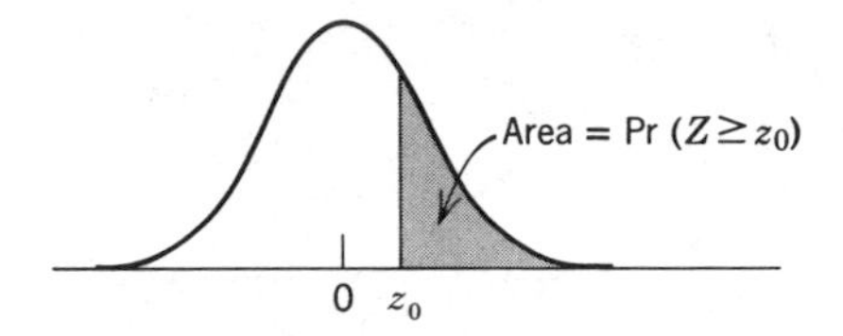

	Second Decimal Place of z_0									
$\rightarrow$ $\downarrow z_0$	.00	.01	.02	.03	.04	.05	.06	.07	.08	.09
0.0	.5000	.4960	.4920	.4880	.4840	.4801	.4761	.4721	.4681	.4641
0.1	.4602	.4562	.4522	.4483	.4443	.4404	.4364	.4325	.4286	.4247
0.2	.4207	.4168	.4129	.4090	.4052	.4013	.3974	.3936	.3897	.3859
0.3	.3821	.3783	.3745	.3707	.3669	.3632	.3594	.3557	.3520	.3483
0.4	.3446	.3409	.3372	.3336	.3300	.3264	.3228	.3192	.3156	.3121
0.5	.3085	.3050	.3015	.2981	.2946	.2912	.2877	.2843	.2810	.2776
0.6	.2743	.2709	.2676	.2643	.2611	.2578	.2546	.2514	.2483	.2451
0.7	.2420	.2389	.2358	.2327	.2296	.2266	.2236	.2206	.2177	.2148
0.8	.2119	.2090	.2061	.2033	.2005	.1977	.1949	.1922	.1894	.1867
0.9	.1841	.1814	.1788	.1762	.1736	.1711	.1685	.1660	.1635	.1611
1.0	.1587	.1562	.1539	.1515	.1492	.1469	.1446	.1423	.1401	.1379
1.1	.1357	.1335	.1314	.1292	.1271	.1251	.1230	.1210	.1190	.1170
1.2	.1151	.1131	.1112	.1093	.1075	.1056	.1038	.1020	.1003	.0985
1.3	.0968	.0951	.0934	.0918	.0901	.0885	.0869	.0853	.0838	.0823
1.4	.0808	.0793	.0778	.0764	.0749	.0735	.0722	.0708	.0694	.0681
1.5	.0668	.0655	.0643	.0630	.0618	.0606	.0594	.0582	.0571	.0559
1.6	.0548	.0537	.0526	.0516	.0505	.0495	.0485	.0475	.0465	.0455
1.7	.0446	.0436	.0427	.0418	.0409	.0401	.0392	.0384	.0375	.0367
1.8	.0359	.0352	.0344	.0336	.0329	.0322	.0314	.0307	.0301	.0294
1.9	.0287	.0281	.0274	.0268	.0262	.0256	.0250	.0244	.0239	.0233
2.0	.0228	.0222	.0217	.0212	.0207	.0202	.0197	.0192	.0188	.0183
2.1	.0179	.0174	.0170	.0166	.0162	.0158	.0154	.0150	.0146	.0143
2.2	.0139	.0136	.0132	.0129	.0125	.0122	.0119	.0116	.0113	.0110
2.3	.0107	.0104	.0102	.0099	.0096	.0094	.0091	.0089	.0087	.0084
2.4	.0082	.0080	.0078	.0075	.0073	.0071	.0069	.0068	.0066	.0064
2.5	.0062	.0060	.0059	.0057	.0055	.0054	.0052	.0051	.0049	.0048
2.6	.0047	.0045	.0044	.0043	.0041	.0040	.0039	.0038	.0037	.0036
2.7	.0035	.0034	.0033	.0032	.0031	.0030	.0029	.0028	.0027	.0026
2.8	.0026	.0025	.0024	.0023	.0023	.0022	.0021	.0021	.0020	.0019
2.9	.0019	.0018	.0017	.0017	.0016	.0016	.0015	.0015	.0014	.0014
3.0	.00135									
3.5	.000 233									
4.0	.000 031 7									
4.5	.000 003 40									
5.0	.000 000 287									

Source: Wonnacott and Wonnacott, *Econometrics*, (1979), p. 538. Copyright © 1979 John Wiley & Sons, Inc. Reprinted with permission.

TABLE B.3. *F* Critical Points

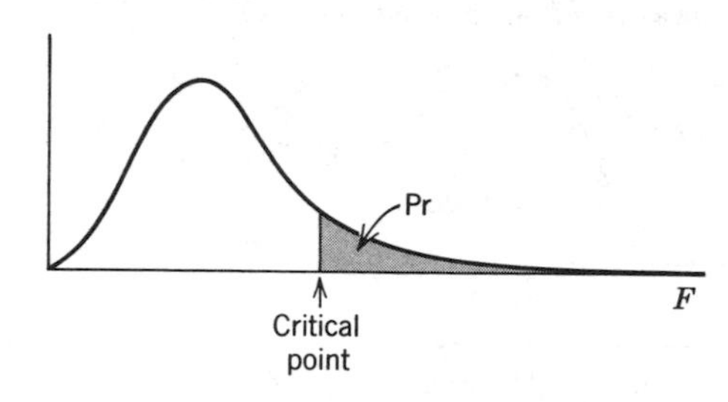

Degrees of freedom for denominator	Pr	Degrees of freedom for numerator 1	2	3	4	5	6	8	10	20	40	∞
1	.25	5.83	7.50	8.20	8.58	8.82	8.98	9.19	9.32	9.58	9.71	9.85
	.10	39.9	49.5	53.6	55.8	57.2	58.2	59.4	60.2	61.7	62.5	63.3
	.05	161	200	216	225	230	234	239	242	248	251	254
2	.25	2.57	3.00	3.15	3.23	3.28	3.31	3.35	3.38	3.43	3.45	3.48
	.10	8.53	9.00	9.16	9.24	9.29	9.33	9.37	9.39	9.44	9.47	9.49
	.05	18.5	19.0	19.2	19.2	19.3	19.3	19.4	19.4	19.4	19.5	19.5
	.01	98.5	99.0	99.2	99.2	99.3	99.3	99.4	99.4	99.4	99.5	99.5
	.001	998	999	999	999	999	999	999	999	999	999	999
3	.25	2.02	2.28	2.36	2.39	2.41	2.42	2.44	2.44	2.46	2.47	2.47
	.10	5.54	5.46	5.39	5.34	5.31	5.28	5.25	5.23	5.18	5.16	5.13
	.05	10.1	9.55	9.28	9.12	9.10	8.94	8.85	8.79	8.66	8.59	8.53
	.01	34.1	30.8	29.5	28.7	28.2	27.9	27.5	27.2	26.7	26.4	26.1
	.001	167	149	141	137	135	133	131	129	126	125	124
4	.25	1.81	2.00	2.05	2.06	2.07	2.08	2.08	2.08	2.08	2.08	2.08
	.10	4.54	4.32	4.19	4.11	4.05	4.01	3.95	3.92	3.84	3.80	3.76
	.05	7.71	6.94	6.59	6.39	6.26	6.16	6.04	5.96	5.80	5.72	5.63
	.01	21.2	18.0	16.7	16.0	15.5	15.2	14.8	14.5	14.0	13.7	13.5
	.001	74.1	61.3	56.2	53.4	51.7	50.5	49.0	48.1	46.1	45.1	44.1
5	.25	1.69	1.85	1.88	1.89	1.89	1.89	1.89	1.89	1.88	1.88	1.87
	.10	4.06	3.78	3.62	3.52	3.45	3.40	3.34	3.30	3.21	3.16	3.10
	.05	6.61	5.79	5.41	5.19	5.05	4.95	4.82	4.74	4.56	4.46	4.36
	.01	16.3	13.3	12.1	11.4	11.0	10.7	10.3	10.1	9.55	9.29	9.02
	.001	47.2	37.1	33.2	31.1	29.8	28.8	27.6	26.9	25.4	24.6	23.8
6	.25	1.62	1.76	1.78	1.79	1.79	1.78	1.77	1.77	1.76	1.75	1.74
	.10	3.78	3.46	3.29	3.18	3.11	3.05	2.98	2.94	2.84	2.78	2.72
	.05	5.99	5.14	4.76	4.53	4.39	4.28	4.15	4.06	3.87	3.77	3.67
	.01	13.7	10.9	9.78	9.15	8.75	8.47	8.10	7.87	7.40	7.14	6.88
	.001	35.5	27.0	23.7	21.9	20.8	20.0	19.0	18.4	17.1	16.4	15.8
7	.25	1.57	1.70	1.72	1.72	1.71	1.71	1.70	1.69	1.67	1.66	1.65
	.10	3.59	3.26	3.07	2.96	2.88	2.83	2.75	2.70	2.59	2.54	2.47
	.05	5.59	4.74	4.35	4.12	3.97	3.87	3.73	3.64	3.44	3.34	3.23
	.01	12.2	9.55	8.45	7.85	7.46	7.19	6.84	6.62	6.16	5.91	5.65
	.001	29.3	21.7	18.8	17.2	16.2	15.5	14.6	14.1	12.9	12.3	11.7
8	.25	1.54	1.66	1.67	1.66	1.66	1.65	1.64	1.63	1.61	1.59	1.58
	.10	3.46	3.11	2.92	2.81	2.73	2.67	2.59	2.54	2.42	2.36	2.29
	.05	5.32	4.46	4.07	3.84	3.69	3.58	3.44	3.35	3.15	3.04	2.93
	.01	11.3	8.65	7.59	7.01	6.63	6.37	6.03	5.81	5.36	5.12	4.86
	.001	25.4	18.5	15.8	14.4	13.5	12.9	12.0	11.5	10.5	9.92	9.33
9	.25	1.51	1.62	1.63	1.63	1.62	1.61	1.60	1.59	1.56	1.55	1.53
	.10	3.36	3.01	2.81	2.69	2.61	2.55	2.47	2.42	2.30	2.23	2.16
	.05	5.12	4.26	3.86	3.63	3.48	3.37	3.23	3.14	2.94	2.83	2.71
	.01	10.6	8.02	6.99	6.42	6.06	5.80	5.47	5.26	4.81	4.57	[illegible]
	.001	22.9	16.4	13.9	12.6	11.7	11.1	10.4	9.89	8.90	8.37	[illegible]

TABLE B.3. *Continued*

Degrees of freedom for denominator	*Pr*	Degrees of freedom for numerator *1*	*2*	*3*	*4*	*5*	*6*	*8*	*10*	*20*	*40*	∞
10	.25	1.49	1.60	1.60	1.59	1.59	1.58	1.56	1.55	1.52	1.51	1.48
	.10	3.28	2.92	2.73	2.61	2.52	2.46	2.38	2.32	2.20	2.13	2.06
	.05	4.96	4.10	3.71	3.48	3.33	3.22	3.07	2.98	2.77	2.66	2.54
	.01	10.0	7.56	6.55	5.99	5.64	5.39	5.06	4.85	4.41	4.17	3.91
	.001	21.0	14.9	12.6	11.3	10.5	9.92	9.20	8.75	7.80	7.30	6.76
12	.25	1.56	1.56	1.56	1.55	1.54	1.53	1.51	1.50	1.47	1.45	1.42
	.10	3.18	2.81	2.61	2.48	2.39	2.33	2.24	2.19	2.06	1.99	1.90
	.05	4.75	3.89	3.49	3.26	3.11	3.00	2.85	2.75	2.54	2.43	2.30
	.01	9.33	6.93	5.95	5.41	5.06	4.82	4.50	4.30	3.86	3.62	3.36
	.001	18.6	13.0	10.8	9.63	8.89	8.38	7.71	7.29	6.40	5.93	5.42
14	.25	1.44	1.53	1.53	1.52	1.51	1.50	1.48	1.46	1.43	1.41	1.38
	.10	3.10	2.73	2.52	2.39	2.31	2.24	2.15	2.10	1.96	1.89	1.80
	.05	4.60	3.74	3.34	3.11	2.96	2.85	2.70	2.60	2.39	2.27	2.13
	.01	8.86	5.51	5.56	5.04	4.69	4.46	4.14	3.94	3.51	3.27	3.00
	.001	17.1	11.8	9.73	8.62	7.92	7.43	6.80	6.40	5.56	5.10	4.60
16	.25	1.42	1.51	1.51	1.50	1.48	1.48	1.46	1.45	1.40	1.37	1.34
	.10	3.05	2.67	2.46	2.33	2.24	2.18	2.09	2.03	1.89	1.81	1.72
	.05	4.49	3.63	3.24	3.01	2.85	2.74	2.59	2.49	2.28	2.15	2.01
	.01	8.53	6.23	5.29	4.77	4.44	4.20	3.89	3.69	3.26	3.02	2.75
	.001	16.1	11.0	9.00	7.94	7.27	6.81	6.19	5.81	4.99	4.54	4.06
18	.25	1.41	1.50	1.49	1.48	1.46	1.45	1.43	1.42	1.38	1.35	1.32
	.10	3.01	2.62	2.42	2.29	2.20	2.13	2.04	1.98	1.84	1.75	1.66
	.05	4.41	3.55	3.16	2.93	2.77	2.66	2.51	2.41	2.19	2.06	1.92
	.01	8.29	6.01	5.09	4.58	4.25	4.01	3.71	3.51	3.08	2.84	2.57
	.001	15.4	10.4	8.49	7.46	6.81	6.35	5.76	5.39	4.59	4.15	3.67
20	.25	1.40	1.49	1.48	1.46	1.45	1.44	1.42	1.40	1.36	1.33	1.29
	.10	2.97	2.59	2.38	2.25	2.16	2.09	2.00	1.94	1.79	1.71	1.61
	.05	4.35	3.49	3.10	2.87	2.71	2.60	2.45	2.35	2.12	1.99	1.84
	.01	8.10	5.85	4.94	4.43	4.10	3.87	3.56	3.37	2.94	2.69	2.42
	.001	14.8	9.95	8.10	7.10	6.46	6.02	5.44	5.08	4.29	3.86	3.38
30	.25	1.38	1.45	1.44	1.42	1.41	1.39	1.37	1.35	1.30	1.27	1.23
	.10	2.88	2.49	2.28	2.14	2.05	1.98	1.88	1.82	1.67	1.57	1.46
	.05	4.17	3.32	2.92	2.69	2.53	2.42	2.27	2.16	1.93	1.79	1.62
	.01	7.56	5.39	4.51	4.02	3.70	3.47	3.17	2.98	2.55	2.30	2.01
	.001	13.3	8.77	7.05	6.12	5.53	5.12	4.58	4.24	3.49	3.07	2.59
40	.25	1.36	1.44	1.42	1.40	1.39	1.37	1.35	1.33	1.28	1.24	1.19
	.10	2.84	2.44	2.23	2.09	2.00	1.93	1.83	1.76	1.61	1.51	1.38
	.05	4.08	3.23	2.84	2.61	2.45	2.34	2.18	2.08	1.84	1.69	1.51
	.01	7.31	5.18	4.31	3.83	3.51	3.29	2.99	2.80	2.37	2.11	1.80
	.001	12.6	8.25	6.60	5.70	5.13	4.73	4.21	3.87	3.15	2.73	2.23
60	.25	1.35	1.42	1.41	1.38	1.37	1.35	1.32	1.30	1.25	1.21	1.15
	.10	2.79	2.39	2.18	2.04	1.95	1.87	1.77	1.71	1.54	1.44	1.29
	.05	4.00	3.15	2.76	2.53	2.37	2.25	2.10	1.99	1.75	1.59	1.39
	.01	7.08	4.98	4.13	3.65	3.34	3.12	2.82	2.63	2.20	1.94	1.60
	.001	12.0	7.76	6.17	5.31	4.76	4.37	3.87	3.54	2.83	2.41	1.89
120	.25	1.34	1.40	1.39	1.37	1.35	1.33	1.30	1.28	1.22	1.18	1.10
	.10	2.75	2.35	2.13	1.99	1.90	1.82	1.72	1.65	1.48	1.37	1.19
	.05	3.92	3.07	2.68	2.45	2.29	2.17	2.02	1.91	1.66	1.50	1.25
	.01	6.85	4.79	3.95	3.48	3.17	2.96	2.66	2.47	2.03	1.76	1.38
	.001	11.4	7.32	5.79	4.95	4.42	4.04	3.55	3.24	2.53	2.11	1.54
∞	.25	1.32	1.39	1.37	1.35	1.33	1.31	1.28	1.25	1.19	1.14	1.00
	.10	2.71	2.30	2.08	1.94	1.85	1.77	1.67	1.60	1.42	1.30	1.00
	.05	3.84	3.00	2.60	2.37	2.21	2.10	1.94	1.83	1.57	1.39	1.00
	.01	6.63	4.61	3.78	3.32	3.02	2.80	2.51	2.32	1.88	1.59	1.00
	.001	10.8	6.91	5.42	4.62	4.10	3.74	3.27	2.96	2.27	1.84	1.00

Source: Wonnacott and Wonnacott, *Econometrics*, (1979), pp. 542–543. Copyright © 1979 John Wiley & Sons, Inc. Reprinted by permission.

TABLE B.4. χ^2 Critical Points

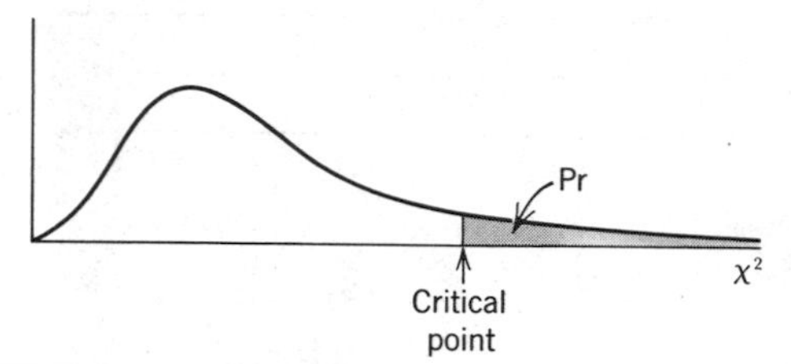

d.f. \ *Pr*	*.250*	*.100*	*.050*	*.025*	*.010*	*.005*	*.001*
1	1.32	2.71	3.84	5.02	6.63	7.88	10.8
2	2.77	4.61	5.99	7.38	9.21	10.6	13.8
3	4.11	6.25	7.81	9.35	11.3	12.8	16.3
4	5.39	7.78	9.49	11.1	13.3	14.9	18.5
5	6.63	9.24	11.1	12.8	15.1	16.7	20.5
6	7.84	10.6	12.6	14.4	16.8	18.5	22.5
7	9.04	12.0	14.1	16.0	18.5	20.3	24.3
8	10.2	13.4	15.5	17.5	20.1	22.0	26.1
9	11.4	14.7	16.9	19.0	21.7	23.6	27.9
10	12.5	16.0	18.3	20.5	23.2	25.2	29.6
11	13.7	17.3	19.7	21.9	24.7	26.8	31.3
12	14.8	18.5	21.0	23.3	26.2	28.3	32.9
13	16.0	19.8	22.4	24.7	27.7	29.8	34.5
14	17.1	21.1	23.7	26.1	29.1	31.3	36.1
15	18.2	22.3	25.0	27.5	30.6	32.8	37.7
16	19.4	23.5	26.3	28.8	32.0	34.3	39.3
17	20.5	24.8	27.6	30.2	33.4	35.7	40.8
18	21.6	26.0	28.9	31.5	34.8	37.2	42.3
19	22.7	27.2	30.1	32.9	36.2	38.6	32.8
20	23.8	28.4	31.4	34.2	37.6	40.0	45.3
21	24.9	29.6	32.7	35.5	38.9	41.4	46.8
22	26.0	30.8	33.9	36.8	40.3	42.8	48.3
23	27.1	32.0	35.2	38.1	41.6	44.2	49.7
24	28.2	33.2	36.4	39.4	32.0	45.6	51.2
25	29.3	34.4	37.7	40.6	44.3	46.9	52.6
26	30.4	35.6	38.9	41.9	45.6	48.3	54.1
27	31.5	36.7	40.1	43.2	47.0	49.6	55.5
28	32.6	37.9	41.3	44.5	48.3	51.0	56.9
29	33.7	39.1	42.6	45.7	49.6	52.3	58.3
30	34.8	40.3	43.8	47.0	50.9	53.7	59.7
40	45.6	51.8	55.8	59.3	63.7	66.8	73.4
50	56.3	63.2	67.5	71.4	76.2	79.5	86.7
60	67.0	74.4	79.1	83.3	88.4	92.0	99.6
70	77.6	85.5	90.5	95.0	100	104	112
80	88.1	96.6	102	107	112	116	125
90	98.6	108	113	118	124	128	137
100	109	118	124	130	136	140	149

Source: Wonnacott and Wonnacott, *Econometrics*, (1979), p. 540. Copyright © 1979 John Wiley & Sons, Inc. Reprinted by permission.

REFERENCES

Allison, P. (1984), *Event History Analysis: Regression for Longitudinal Data*. Beverly Hills: Sage Publications.

Amemiya, T. (1981), "Qualitative response models: A survey," *Journal of Economic Literature*, Vol. 19, pp. 1483–1538.

Andrews, F., J. Morgan, and J. Sonquist (1969), *Multiple Classification Analysis*. Ann Arbor: Survey Research Center, Institute for Social Research, University of Michigan.

Asher, H. B. (1983), *Causal Modeling*. Beverly Hills: Sage Publications.

Barclay, G. W. (1958), *Techniques of Population Analysis*. New York: Wiley.

Benedetti, J., K. Yuen, and L. Young (1981), "Life tables and survival functions," in W. J. Dixon (ed.) (1981).

Blau, P. M. and O. D. Duncan (1967), *The American Occupational Structure*. New York: Wiley.

Breslow, N. E. (1970), "A generalized Kruskal–Wallis test for comparing K samples subject to unequal patterns of censorship," *Biometrika*, Vol. 57, pp. 579–594.

Chiang, C. L. (1968), *Introduction to Stochastic Processes in Biostatistics*. New York: Wiley.

Cochran, W. G. (1963), *Sampling Techniques*, second edition. New York: Wiley.

Cox, D. R. (1970), *The Analysis of Binary Data*. London: Methuen.

Cox, D. R. (1972), "Regression models and life tables," *Journal of the Royal Statistical Society*, *B*, Vol. 34, pp. 187–220.

Cox, D. R. and D. Oakes (1984), *Analysis of Survival Data*. London: Chapman and Hall.

Cragg, J. G. and R. Uhler (1970), "The demand for automobiles," *Canadian Journal of Economics*, Vol. 3, pp. 386–406.

Dixon, W. J. (ed.) (1981), *BMDP Statistical Software 1981*. Berkeley: University of California Press.

Dixon, W. J. (ed.) (1992), *BMDP Statistical Software Manual* (in two volumes), Berkeley: University of California Press.

Duncan, O. D. (1975), *Introduction to Structural Equation Models*. New York: Academic Press.

Fox, J. (1984), *Linear Statistical Models and Related Methods*. New York: Wiley.

Gehan, E. A. (1965), "A generalized Wilcoxon test for comparing arbitrarily singly-censored samples," *Biometrika*, Vol. 52, pp. 203–223.

Green, W. H. (1992), LIMDEP User's Manual and Reference Guide, Version 6.0. Bellport, N.Y.: Economic Software, Inc.

Hanushek, E. J. and J. E. Jackson (1977), *Statistical Methods for Social Scientists*. Orlando: Academic Press.

Harrell, F. E. (1983), "The LOGIST Procedure," pp. 181–202 in *SUGI Supplemental Library User's Guide, 1983 Edition*. Cary, North Carolina: SAS Institute Inc.

Holland, P. W. (1986), "Statistics and causal inference," *Journal of the American Statistical Association*, Vol. 81, pp. 945–960.

Holland, P. W. (1988), "Causal inference, path analysis, and recursive structural equations models," pp. 449–484 in C. C. Clogg (ed.), *Sociological Methodology 1988*. Washington, D.C.: American Sociological Association.

Kalbfleisch, J. D. and R. L. Prentice (1980), *The Structural Analysis of Failure Time Data*. New York: Wiley.

Kendall, M. G. and C. A. O'Muircheartaigh (1977), *Path Analysis and Model Building*. World Fertility Survey Technical Bulletin No. 2. The Hague: International Statistical Institute.

Maddala, G. S. (1983), *Limited-Dependent and Qualitative Variables in Econometrics*. Cambridge: Cambridge University Press.

Mantel, N. (1966), "Evaluation of survival data and two new rank order statistics arising in its consideration," *Cancer Chemotherapy Reports*, Vol. 50, pp. 163–170.

McFadden, D. (1974), The measurement of urban travel demand," *Journal of Public Economics*, Vol. 3, pp. 303–328.

Mood, A. M., F. A. Graybill, and D. C. Boes (1974), *Introduction to the Theory of Statistics*. New York: McGraw–Hill.

Morrison, D. G. (1972), "Upper bounds for correlations between binary outcomes and probabilistic predictions," *Journal of the American Statistical Association*, Vol. 67, pp. 68–70.

SAS Institute (1990), *SAS/STAT User's Guide, Version 6*, fourth edition, Volumes 1 and 2. Cary, NC: SAS Institute.

Schmidt, P. and R. P. Strauss (1975), "The predictions of occupation using multiple logit models," *International Economic Review*, Vol. 17, pp. 204–212.

Stuart, A. and J. K. Ord (1987 and 1991, in two volumes), *Kendall's Advanced Theory of Statistics*. New York: Oxford University Press.

Theil, H. (1969), "A multinomial extension of the linear logit model," *International Economic Review*, Vol. 10, pp. 251–259.

Theil, H. (1971), *Principles of Econometrics*. New York: Wiley.

Walker, S. H. and D. B. Duncan (1967), "Estimation of the probability of an event as a function of several independent variables," *Biometrika*, Vol. 54, pp. 167–179.

Wonnacott, R. J. and T. H. Wonnacott (1979), *Econometrics*. New York: Wiley.

Wonnacott, T. H. and R. J. Wonnacott (1984), *Introductory Statistics for Business and Economics*. New York: Wiley.

INDEX

Actuarial life table, 168–175
ANOVA table, 59–60, 66
Additive model, 33, 42, 132, 189. *See also* Models
Adjusted effects (in multiple classification analysis), 74
Adjusted *R*, 75. *See also* Coefficient of determination
Adjusted values (in multiple classification analysis), 70, 73
Alternative hypothesis, 13
Analysis:
  of covariance, 69
  of variance, 59, 69
Assumptions, regression, 8, 29
  homoscedasticity, 9, 29
  independence, 9, 29
  linearity, 8, 29
  sampling, 9, 30

Backward deletion, 62
Baseline hazard, 182, 185, 209
Beta (in multiple classification analysis), 76
Bias, 10, 30, 31
Binary variable, 34, 119. *See also* Variables
Bivariate linear regression, 1–28
BMDP, 21, 121, 142, 230–238
Breslow test, 177. *See also* Tests of significance

Categorical variable, 2, 34, 36, 151. *See also* Variables
Causality, *see also* Effects
  causal model, 5–7
  causal ordering, 94
  path analysis, 93–118
  simultaneity, 6
  two-way causation, 6
Censoring:
  censored observation, 171
  censoring time, 175
  adjustments for, 171–172
Chi square (χ^2):
  critical points, 250
  distribution, 250
  test, 139, 177. *See also* Tests of significance
Coefficients:
  bivariate regression, 2
  multiple regression, 33
  partial regression, 33
  path, 94–109
Coefficient of determination:
  adjusted R^2, 75
  corrected R^2, 53
  R^2, 52
  r^2, 26
Cohort, definition of, 167
Compound path, 96. *See also* Path analysis
Conditional probability, 168. *See also* Probability
Confidence interval, 15, 55. *See also* Tests of significance
Continuous variable, 2. *See also* Variables
Contrast variable, 34. *See also* Variables
Control variable, 33, 34, 73, 85, 101, 105. *See also* Variables
Corrected R^2, 53. *See also* Coefficient of determination
Correlation, *see also* Coefficient of determination

correlation coefficient, 26
correlation matrix, 40
correlation without interaction, 45
interaction without correlation, 45
multiple correlation coefficient, 52
partial correlation, 54–55
Covariance, *see also* Analysis
population, 27
regression coefficient covariance matrix, 57
sample, 27
Critical points, 17. *See also* Tests of significance
Critical values, 17. *See also* Tests of significance
Cumulative hazard, 179

Degrees of freedom (d.f.), 12, 56, 57, 58, 59,
Dependent variable, 1. *See also* Variables
Deviation form (in multiple classification analysis), 78–82
Deviation form (in proportional hazard models), 186
Dichotomous variable, 34, 119. *See also* Variables
Direct effect, 96. *See also* Effects
Direct path, 96
Distribution-dependent effects, 113. *See also* Effects
Disturbance, in regression, 9, 30
Dummy variable, 34, 36. *See also* Variables

Effects:
additive effect, 5, 27, 33, 37, 128, 153, 188
direct effect, 96
distribution-dependent effect, 113
indirect effect, 96
interaction effect, 41, 44, 57, 135, 159, 191
main effect, 41
multiplicative effect, 131, 153, 189, 209
nonlinear effect, 46, 48, 58, 136, 159, 192
time-dependent effect, 214
total effect, 96
Endogenous variable, 1, 94. *See also* Variables
Error, in regression, 9, 30
Error sum of squares, 25. *See also* Sum of squares
Estimated model, 9, 30. *See also* Models
Estimator, 7
Eta (in multiple classification analysis), 76
Exogenous variable, 1, 94. *See also* Variables
Explained sum of squares (ESS), 25, 52. *See also* Sum of squares
Explained variation, *see* Explained sum of squares
Explanatory variable, 1. *See also* Variables

Factor, 2. *See also* Variables
Failure, definition of, 167
Failure time, 175
Feedback loop, 95
F critical points, 248
F distribution, 248
F ratio, 60
F test, 60, 61. *See also* Tests of significance

Generalized Gehan test, 177. *See also* Tests of significance
Generalized Savage test, 177. *See also* Tests of significance
Generalized Wilcoxon test, 177. *See also* Tests of significance
Goodness of fit, 23, 27, 51, 140, 157, 198
Grand mean, 78

Hazard function, 172, 182, 185
Hazard model, 181–220. *See also* Models
in deviation form, 186
proportional hazard model, 181-206
in survivorship form, 194
with time dependence, 207–220
Hazard rate, 172
Heterogeneity, 173, 181, 195
Homogeneity, 173, 181, 195
Homoscedasticity, 9. *See also* Assumptions, regression
Hypothesis testing, *see* Tests of significance

"Improper" odds, 152
Independence, statistical, 9. *See also* Assumptions, regression
Independent variable, 1, 30. *See also* Variables
Indirect effect, 96. *See also* Effects
Indirect path, 96. *See also* Path analysis
Inflection point, 123
Instantaneous rate of change, 49
Interaction, 40–44, 57, 82, 114, 135, 159, 191
correlation without interaction, 45
interaction effect, 41. *See also* Effects
interaction without correlation, 45
interactive model, 42. *See also* Models
three-way, 44
Intercept, 2,5
Intercept model, 140. *See also* Test model
Interval estimate, 16
Intervening variable, 96. *See also* Variables

Kaplan–Meier life table, *see* Product-limit life table

Life table, 167–180
  actuarial life table, 168–175
  life table calculated from proportional hazard model, 194
  multivariate life table, 181, 194–195
  product-limit (Kaplan–Meier) life table, 175–178
Likelihood:
  L, 138
  likelihood function, 147
  likelihood ratio test, 139
  log likelihood, 138, 149
  maximum likelihood method, 147–150
LIMDEP, 21, 121, 142, 238–244
Linear probability model, 119–121. *See also* Models
Linearity:
  in the coefficients, 48
  in the variables, 48
Linearity assumption in regression, 8, 29. *See also* Assumptions, regression
Log hazard, 188
Log likelihood, 138, 149
Log odds, *see* Logit
Logistic function, 122–126
Logistic regression, *see* Logit regression
Logit, definition of, 127
Logit regression, 119–150. *See also* Regression

Main effect, 41. *See also* Effects
Mantel test, 177. *See also* Tests of significance
Mantel–Cox test, 177. *See also* Tests of significance
Maximum likelihood method, 147–150. *See also* Likelihood
Mean squared error (MSE), 24
Model(s):
  additive, 33, 42, 132, 189
  bivariate linear regression, 1–28
  definition of a model, 2
  estimated, 9, 30
  hazard model with time dependence, 207–220
  interactive, 42–44, 83
  intercept model, 140, 157
  linear probability model, 119–121
  logit regression, 121–128
  multinomial logit regression, 151–165
  multiple regression, 29–68
  multiplicative, 131, 132, 189
  nested, 138
  nonrecursive, 95
  path, 93
  population model, 9, 30
  proportional hazard, 181–206
  recursive, 95
  statistical model, 7
  test model, 140
Model specification, 2, 40
Model sum of squares, 25. *See also* Sum of squares
Multicollinearity, 38
Multinomial logit model, 151–165
Multiple classification analysis (MCA):
  hazard model with time dependence, 151–165
  logit regression, 142–147
  multinomial logit regression, 153–157
  multiple regression, 69–92
  proportional hazard regression, 205–206
Multiple correlation coefficient, 52. *See also* Correlation
Multiple regression, 29–68
Multiplicative model, 131, 132, 189. *See also* Models
Multivariate analysis, 1
Multivariate life table, 181, 194–195
Multivariate logistic function, 125

Nested model, 138. *See also* Models
Nonlinearity:
  in the coefficients, 48
  in the variables, 46, 114, 136, 159, 192
Nonrecursive model, 95. *See also* Models
Normal approximation, 11
Normal distribution, 11, 12, 247
Normal equations, 4, 32
Null hypothesis, 12

Observed level of significance, *see p* value
Odds, 126–127
  odds, improper, 152
Odds ratio, 131
Omitted category, *see* Reference category
Omnibus *F* test, 60. *See also* Tests of significance
One-tailed tests, *see* Tests of significance
One-way analysis of variance, 69
Ordinary least squares (OLS), 4
Outliers, 20

Parabola, 46–47
Parity progression ratio, 172–173
Partial correlation coefficient, 54–55. *See also* Correlation
Partial likelihood function, 209
Partial *R*, 54–55
Partial *r*, 54
Partial regression coefficient, 54–55. *See also* Coefficients

Partially reduced form, 97
Path analysis, 93–118
  causal diagram, 93–94
  compound path, 96
  direct path, 96
  indirect path, 96
  path coefficients, 94
  path diagram, 93–94
  path model, 93
  residual path, 111–113
  simple path, 96
Prob value, *see* p value
Piecewise-linear, 49
Point estimate, 16
Polytomous logit model, *see* Multinomial logit model
Population model, 30. *See also* Models
Predicted value, 3
Predictor variable, 1. *See also* Variables
Probability:
  conditional probability of failure, 168
  conditional probability of survival, 168
  cumulative probability of failure, 168
  cumulative probability of survival, 168
  probability in logit regression, 123, 125, 134
  probability in multinomial logit regression, 151, 154
Prob value, *see* p value
Product-limit life table, 175–177
Proportional hazard(s) model, 181–206. *See also* Models
Proportionality assumption in hazard model, 182–183, 192, 193
Pseudo-R, 181, 157, 198
p value, 17, 55

Quadratic term, 46
Qualitative variable, 2. *See also* Variables
Quantitative variable, 2. *See also* Variables

Recursive model, 95. *See also* Models
Reduced form, 95, 97
Reference category, 35, 36, 37
  of the response variable, 157
Regression:
  assumptions, 8, 29
  bivariate linear regression, 1–28
  hazard regression with time dependence, 207–220
  logit regression, 119–150
  multinomial logit regression, 151–165
  multiple regression, 29–68
  proportional hazard regression, 181–206
  stepwise regression, 62
Regression coefficient, *see* Coefficients
Regression coefficient covariance matrix, 57
Regression sum of squares, 25. *See also* Sum of squares
Regressor, 1
Relative risk, 187–188
Residual, in regression, 3, 30
Residual path coefficient, 112
Residual sum of squares, 25. *See also* Sum of squares
Residual variable, in path analysis, 110, 112. *See also* Variables
Residual variance, 112
Response variable, 1. *See also* Variables
Robustness, 10
Rounding errors, 204–205

SAS, 21, 121, 223–229
Sample covariance, 27. *See also* Covariance
Sample point, 3
Sample standard deviation, *see* Standard deviation
Sampling, 9, 10. *See also* Assumptions, regression
Saturated model, 107
Scatterplot, 21
Selection on response variable, 21
Sigmoid curve, 121
Significance, level of, 14, 17
Significance, statistical, *see* Tests of significance
Significant digits, 204
Simple path, 96. *See also* Path analysis
Simultaneity, 6. *See also* Causality
Simultaneous hypothesis test, 60. *See also* Tests of significance
Specification, *see* Model specification
Specification error, 2, 31
SPSS, 21, 121
Standard deviation:
  population, 27
  sample, 27
Standard error, 11, 55
  of the estimate, 24, 51
Standardized path coefficient, 109, 113
Standardized regression coefficient, 109, 113
Standardized variable, 109. *See also* Variables
Standard normal probabilities, 247
Stepwise regression, 62
Stratification variable, 22. *See also* Variables
Stratified sample, 22
Structural equation model, 95
Student's t distribution, *see* t distribution
Sum of squares, *see also* Variation
  error sum of squares, 25
  explained sum of squares (ESS), 25, 52

Sum of squares (*Continued*)
  model sum of squares, 25
  regression sum of squares, 25
  residual sum of squares, 25
  total sum of squares (TSS), 25
  unexplained sum of squares (USS), 25, 52
Survival function, 168, 175, 216
Survival models, 167–220
Survival time, 175

t distribution, 11, 12, 246
Test for proportionality, 196
Tests of significance:
  Breslow test, 177
  chi square (χ^2) test, 139
  confidence intervals, 15
  critical points, 17
  critical values, 17
  difference between two coefficients, 56
  difference between two models, 138
  difference between two survival curves, 177
  F test that all coefficients are zero, 60
  F test that some coefficients are zero, 61
  generalized Gehan test, 177
  generalized Savage test, 177
  generalized Wilcoxon test, 177
  likelihood ratio test:
    hazard regression, 198
    logit regression, 138–140
    multinomial logit regression, 157
  Mantel test, 177
  Mantel-Cox test, 177
  omnibus *F* test, 60
  one-tailed test, 13, 55
  simultaneous hypothesis test, 60
  tests when interaction is present, 57, 63–68, 198–205
  tests when nonlinearities are present, 58, 63–68, 198–205
  two-tailed test, 13, 55
Test model, 140. *See also* Models
Theory, definition of, 5
Time-dependent coefficient, 214
Time-dependent predictor variable, 207. *See also* Variables
Total effect, 96. *See also* Effects
Total sum of squares (TSS), 25. *See also* Sum of squares
Total variation, *see* Total sum of squares
t critical points, 246
t distribution, 246
t statistic, 11, 16, 246
Two-tailed test, *see* Tests of significance
Two-way causation, 6. *See also* Causality

Unadjusted effects (in multiple classification analysis), 72
Unadjusted *R* (in multiple classification analysis), 75
Unadjusted values (in multiple classification analysis), 70
Unbiased estimator, 10
Unconditional probability, *see* Probability
Unexplained sum of squares (USS), 25, 52. *See also* Sum of squares
Unexplained variation, *see* Unexplained sum of squares (USS)
Unobserved heterogeneity, 181, 195
Unsaturated model, 107

Variables:
  binary, 34, 119
  categorical, 2, 34, 36, 151
  continuous, 2
  contrast, 34
  control, 33, 34, 73, 85
  dependent, 1
  dichotomous, 34, 119
  dummy, 34, 36
  endogenous, 1, 94
  exogenous, 1, 94
  explanatory, 1
  factor, 2
  independent, 1, 30
  intervening, 96
  predictor, 1
  qualitative, 2
  quantitative, 2
  residual, 110, 112
  response, 1
  standardized, 109
  stratification variable, 22
  time-dependent predictor variable, 207
Variance 59, 69. *See also* Analysis
Variation, *see* Sum of squares

Z score, *see Z* value
Z value, 16, 137, 247